WOOD, WHISK

'Work has done a fine job directing the spotlight toward an object that seems to beg for inattention. Although much diminished from their peak a century ago, coopers are today thriving again with American bourbon makers clamoring for new casks. (Federal regulations require that anything labeled bourbon be aged in new oak casks). The number of craft spirits producers has also surged in the past decade, and barrels are suddenly in short supply. Among vintners, high-quality barrels also remain in high demand, although makers of cheaper wines have embraced workarounds, including the use of oak chips and short planks placed in stainless steel tanks. Work offers a breezy tour through all this and more. When you reach the end of this book, I can pretty much guarantee you won't think of barrels the same way again. Next time you pass a geranium planter made from an old whiskey barrel cleaved in two, take a moment to pause and pay your respects. This was the container that built America.' – *Wall Street Journal*

'There is plenty in the book of interest, and not just for the many fans of wine and whiskey . . . Henry Work sets out to demonstrate the technological, cultural and economic importance of barrels from their development, probably before 500 BC, to their ubiquity for storage and transport from the Middle Ages until the early 20th century.' – *The Spectator*

'A thorough and entertaining journey from amphorae, barrels' predecessors, through their period of domination to their relative demise due to replacement with plastic and metal containers . . . there is much to interest both the general reader and the beer enthusiast in this well-written history of a container that has been with humanity for so long.' – *London Drinker*

'Work, an American cooper now living in New Zealand, writes intimately of his metier and materials, starting with early evidence of the barrel's invention among the Celts . . . Sections on how staves are made – often sawn when the material is American oak, or split when it's European – feel like lessons from a master craftsman.' – TLS

'Henry H. Work, a cooper himself since the 70s, takes us through the two-millennia-long story of cooperage from the birth of the trade to the evolution from bucket to barrel to its function in beverage to the uncertain future of the craft. This is an interesting and thorough look at the modest container's significant role in history. It's sure to give any beer, wine, or whiskey enthusiast a serious (metaphorical) rager – *Craft Magazine* (US)

'An enlightening study of this humble wooden receptacle. In simple, non-academic prose, Work traces the wooden barrel from its Celtic roots, through its heyday as a necessity for seafaring industries, to its current utility in aging alcohol . . . Casual and concise, this is a book for every wine drinker who enjoys a bit of history.' – *Terroirist.com*

'Someone was kind enough to give me a terrific book at Christmas called *Wood, Whiskey and Wine: A History of Barrels* . . . Now it might not look like, yes, a barrel of laughs, but the story of the barrel is a truly fascinating one.' – *Irish Post*

# WOOD, WHISKEY

# Wine

## A HISTORY OF BARRELS

Henry H. Work

REAKTION BOOKS

Published by
Reaktion Books Ltd
Unit 32, Waterside
44–48 Wharf Road
London N1 7UX, UK
www.reaktionbooks.co.uk

First published 2014
First published in paperback 2024

Printed and bound in Great Britain
by Clays Ltd, Elcograf S.p.A.

A catalogue record for this book is available from the British Library

ISBN 978 1 78914 920 3

# Contents

Etching by Leonard Beaumont of Spanish coopers fabricating barrels, *c.* 1930.

# Introduction

For over 2,000 years wooden barrels have been used as bulk containers. Since their initial development, crafted by the Celtic tribes of northern Europe in the first millennium BC, they have stored, transported and aged an incredibly diverse array of provisions and liquids. And despite having been largely replaced by plastic, cardboard and metal, they are still highly valued for ageing the world's finest wines, bourbons and whiskies.

The *pipes*, or port barrels, of Vila Nova de Gaia provide an iconic example of this manifold usage. At this Portuguese town, located opposite Oporto on the Douro River, the barrels can be seen lashed to the decks of the *barcos rabelos*, the flat-bottomed sailing vessels that transported them up and down the river. A view from Google Earth clearly shows several of the boats moored at the quay in front of the red-tiled port lodges. Aboard each boat are six large *pipes*.

Oporto and Vila Nova de Gaia were established towns well before the Romans expanded into what is now Portugal. By the time the legions arrived, the indigenous Celts had rudimentary barrels for the ageing and storage of their wines and ports, plus river-going vessels with which to transport these liquids between the vineyards and cities.[1] Historically, the newly fermented port, from wineries located on the terraced slopes in the upper reaches of the Douro, was shipped down to the ageing and export warehouses at Vila Nova de Gaia on the boats. Today, these *barcos rabelos* are moored along the waterfront primarily for the ambiance; their function was handed over to trucks in the 1960s when the river was dammed.[2]

Ambiance! Years ago the port would be aged and shipped to London or New York still in the barrel. Now, while the port is aged in the pipes, it is disbursed to customers around the world in bottles packed in cardboard cases: a poignant indicator of the way the wooden barrel is slowly being replaced.

Thinking back, do you recall seeing barrels used for products other than beer, wine or whiskey? While barrels were ever present in our grandparents' day, a small number of our parents, and even fewer of us, relied upon them for our provisions.

Growing up on America's East Coast, I recall only a few examples of wooden barrels in use. I do remember that in houses under construction there were kegs containing the bulk nails used by the carpenters. If the workers were friendly, they would flip over an empty keg and allow me to sit on it, listening to their banter as they ate their lunch. These barrels were relatively small, perhaps 40–50 centimetres tall and 20–30 centimetres in diameter. Some were held together by wire and others by thin metal hoops. If the kegs survived the house construction, they might have been used later to hold firewood, golf clubs, hockey sticks or other items in the household garage or basement. Today, nails come in cardboard or plastic boxes, which are generally tossed out when the nails have been used.

Many houses in the eastern United States had a cellar or basement. Within these cellars, some residents cleared their accumulated and unused household goods to make a living space complete with a small bar. One day, while visiting a neighbour's basement where there was such a bar, I saw, proudly sitting on small stands, several kegs which held brandy or port. Some were unadorned, simple kegs, but others, usually those imported from Europe, were made in an oval shape, with the front painstakingly carved. The liquid was dispensed via a wooden or brass spigot from the head of the keg. Nowadays, most of us pour our port or sherry from glass decanters or bottles.

I also remember that wooden barrels were part of the display seen during a visit with my parents to a historic sailing ship moored in the seaport of Mystic, Connecticut. These barrels would have contained the bewildering range of supplies necessary for long ocean voyages, as well as some of the cargo the ship would be carrying for trade. Today, those supplies are stored in plastic bins and boxes, and metal shipping containers are far more commonly used for transporting the cargo. On another occasion, at a sweetshop, I chose my selection from an array displayed in a wooden barrel. This would have been the forerunner to presentations within barrels in stores and restaurants now peddling folksy nostalgia.

By the 1970s and '80s, I saw wooden barrels which had been cut in half to make planters and occasionally whole ones used to catch rainwater, placed at the bottom of a downspout. Although by this time, having moved to California, they probably weren't as plentiful as they were in the eastern states, nearer to the source – the surplus of used whiskey barrels coming from the distilleries in Kentucky and Tennessee. The barrels I saw in California were more likely discarded from the growing premium-wine industry.

But it was when I moved to the Napa Valley in 1974, with its wine fame gaining on that of France's Bordeaux and Burgundy, that my encounters with the wooden barrels become significant and personal. These were the critical ageing vessels for the expanding premium-winemaking industry in California, and so a source of livelihood for me.

Hired by a struggling barrel sales-and-repair enterprise which was exploring all manner of income sources, including making redwood hot tubs – this was, after all, California in the 1970s – I gravitated towards the assignments that involved wine barrels. There was something captivating about the barrels that the Valley's wineries were purchasing in ever-increasing numbers. The historical development and use of wooden barrels begged to be explored. It seemed they would be around longer than wooden hot tubs.

My trial by fire came when two co-workers and I were sent, in the middle of winter, to a large, upstate New York winery. There, we were to assemble several hundred puncheons, or 600-litre barrels. The wood for these puncheons was American oak, which had been shipped to Portugal where, with inexpensive labour, it was made into the barrels. When the barrels were completed, the staves, hoops and heads were carefully numbered. They were then dismantled, to conserve space in the shipping container, bundled and shipped back to the United States to this winery. Our job was to sort through the 200 or so potential barrels and assemble all the pieces correctly; each head was made, and each hoop was sized, specifically for a certain barrel. Ironically, it was wooden barrels that were the 'shipping containers' for the world up until just a relatively few years ago.

Being California boys, we were not acclimatized, or prepared, for the bitter cold in an unheated New York warehouse. The job was further complicated by the height of the puncheons – 1.4 metres tall – and the force needed to work them. These barrels were encircled by heavy metal hoops, and to cinch a barrel together, the metal hoops need to be forced towards the bilge, or middle of the barrel. This is accomplished using a hoop driver: a handheld piece of metal upon which to transfer to the hoop the downward force from a heavy hammer. The action is similar to hammering on a chisel, but instead of light taps, the cooper puts all his strength into bringing the hammer down for each blow. As novice coopers, it was this force that we lacked – we didn't have the strength to swing the hammers high above our shoulders for the many repeated hits necessary to drive down the big hoops. What was anticipated as a one- or two-week job turned into three; and only during the last week did the cursing and sore muscles subside into sustained and productive barrel assembly.

Upon returning to California, while not anxious to repeat my New York experience, I was grateful that it had offered a glimpse not only of a different aspect of the wine industry compared to the practices that I saw in the Napa Valley, but also of the larger world of cooperage – the general term for all wooden pails, buckets, barrels and tanks. That oak wood could be sourced in the United States, coopered in Portugal and then returned to age New York wine was intriguing. Little did I realize, the port barrels resting in the port warehouses at Vila Nova de Gaia and on the *barcos rabelos* at the quay were also made from American oak.

Where else was cooperage traded? How were other wine barrels made, and where? Who made the first barrels, and what were the other uses of barrels through the ages? All these questions piqued my curiosity and imagination. The subsequent years, employed within the cooperage business and including a stint as general manager of a Kentucky bourbon barrel plant which was transformed to craft wine barrels, provided opportunities to search for answers. The hands-on experiences I encountered, with both many various-sized barrels and wooden tanks, coupled with travel to numerous wine and whiskey regions, slowly provided insight into the fascinating world of cooperage.

As cultural icons for the port region, the barrels, resting within the *barcos rabelos*, provide a picturesque reminder of their importance to the port trade and to this area's way of life. And while wooden barrels are displayed within other historic ships in waterfront cities around the globe, few exemplify so well the critical integration between wooden boats and wooden barrels: a fundamental connection in the long history of cooperage.

Wooden barrels, and the associated cooperage containers, have aided the trade and storage of all manner of foodstuffs and supplies, on land and at sea, and contributed to the advancement of most of the water's edge cities of the Western world. These topics, along with how barrels are crafted and used within today's wine and whiskey industries, will be explored throughout this book. Additionally, we will examine the factors which influenced decisions by vintners, distillers, brewers, sailors, fishermen and merchants to contain their products within those vessels.

## A Note on Terminology and Measurement

While the past development of the wooden barrel was a rather unscientific process, to attempt to bring cooperage into the twenty-first century, metric measurement has been used throughout the book. However, where use of the historical measurement makes more sense, I have utilized it, followed by the metric measurement. Gallons are U.S. ones, not British, and currency is denoted in U.S. dollars.

Additionally, 'whisky' will be spelled without the 'e' when referring to Scotch, the spirit produced in Scotland, whereas it will be spelt 'whiskey' when referring to bourbon and generic whiskies.

# ONE

# Need: Why Wooden Barrels?

Wooden barrels have been arguably the most significant shipping
container in history. They served Romans, explorers, pilgrims,
pirates, pioneers and samurai through 2,000 years of civilization.

Diana Twede, 'The Cask Age: The Technology and History of Wooden Barrels'

 The now-ubiquitous plastic water bottle is symptomatic
of our throwaway society. It is also a symbol of 'individualized packaging' and is basically the antithesis of the
barrel. Wooden barrels were, and are, bulk containers. Travellers of
150 years ago on a sailing ship, a steam train or a horse-drawn coach
would have filled their own glass bottles or leather canteens from
water stored in a wooden barrel. On sailing ships, when not on duty,
sailors often hung around the water-storage barrel to get a drink
and to chat (a forerunner of our modern-day sessions at the office
water cooler). The barrel around which they congregated was usually placed on the deck of a sailing ship, near a hatch that was known
in naval terms as a scuttle. Most barrel sizes have specific names,
and this large barrel was called a butt. The conversation, or scuttle-
butt, that ensued took its name from the location and the barrel.
Though the barrel is long gone, the term lives on as a word for gossip.

Barrels were also the container of choice for an extremely wide
range of other commodities. Before the use of individual pockets in
cardboard or plastic trays, apples were packed in what were called
slack barrels – that is, wooden barrels which were not liquid-tight.
Bulk nails and gunpowder were stored in small barrels called kegs.
Because nails needed little protection from the elements, they would
have been packaged in 'slack' kegs, while the gunpowder was kept
dry by being sealed in a 'tight' barrel, the general term for watertight cooperage. Salted meat was packed in hogsheads – capacious
340-litre barrels which were slightly larger than the typical wine or
bourbon barrel. Over the centuries, an enormous range of products

were transported and stored in barrels: animal hides and skins, beer, cement, cheese rennet,[1] cider, coconut oil, coins, cornmeal, crackers, flour, grains, green ginger, molasses, palm oil, paint, petroleum products, pickles, potatoes, putty, salt, salted fish,[2] salted meats, seeds, sugar, syrup, tar, tobacco, vinegar, whale oil, whiskey, wine and even linens and crockery, cushioned in straw.[3] The barrels for these diverse commodities were all roughly similar in shape – the bulbous cylinder with flat ends – but varied considerably in size and in the type of wood used for their construction. Most acquired their own specific names and identities.

In the eighteenth and nineteenth centuries, portions of America's bountiful cod, harvested off the New England coasts, were dried, salted and packed in barrels to be sent to England and Europe.[4] At the height of the United States' Chesapeake Bay oyster harvest, barrels of oysters would be shipped across the country by railroad, and many a household in the Midwest or on the Pacific coast had a barrel of Chesapeake oysters in the cellar.[5] Nineteenth- and twentieth-century French homes may also have had a barrel of oysters, plucked from the cold Atlantic waters, in their cellars and would certainly have

An 18th-century advertisement for Sancho's Best Trinidado tobacco paper, with a tobacco barrel in the background.

13

A postcard of 1907 mailed from Rochester, Kent, showing the hooks used to grab barrels to load or unload them from ships and barges.

had a barrel or two of wine, quietly ageing. In Britain a pint at the local pub would have been drawn from a barrel.

From the time of the early Roman Empire until the early twentieth century, a number of factors meant that the wooden barrel was favoured as the bulk container of choice.[6] Before aluminium and plastic, wooden barrels provided a watertight container. Before tin and steel, wooden barrels offered protection against the depredations of rats. Before collapsible cardboard boxes, wooden barrels could be shipped unassembled, or 'knocked-down', and then assembled prior to packing or filling. And when unused, they could be knocked-down again to await reuse. Before forklifts, barrels containing heavy products, such as water or salted meat, could be easily rolled up ramps into ships or wagons. The barrels nested tightly against one another in the holds of ships or, when turned upright, remained relatively steady throughout the journey.

In the late 1850s, the first wells were successfully drilled to extract the petroleum seeping from western Pennsylvania's sedimentary formations. Wooden barrels of 42-gallons (159-litre) capacity were used to transport that crude oil from the wellhead to the refineries for conversion into oil for heating and lighting. As the need for oil as fuel overtook its original medicinal uses, millions of these barrels were employed.[7] Today, the thousands of tons of oil resting in the vast holds of a supertanker steaming through the Arabian Gulf are still measured in barrels, though the original wooden containers have long since vanished. And that tanker's capacity is measured in 'tons', from the French word *tuns*, meaning large barrels or casks. Long before twist tops on beer bottles or pull tabs on aluminium beer cans, beer was aged, transported and dispensed from barrels. Pork meat used to be transported in barrels too; the term 'pork barrel' now implies, in the U.S. at least, the transportation of money to a Congressman's home district. Barrels were also used to transport coins, and if one was 'scraping the bottom of the barrel' it was to find any remaining cash. And before the common use of glass for wine bottles, wine was transported and stored in barrels, then dispensed though a spigot in the head, or end, of a barrel into a glass, bottle or pitcher.

Barrels had significant advantages over ceramic containers for size and durability. Wooden barrels and tanks could be made much larger without adding significantly to their weight. And they were also less likely to break if dropped or jostled. In the vast forests of Europe, Russia and North America, wood – especially oak, poplar, elm, fir and chestnut – was a plentiful resource. And the use of barrels increased as these and further advantages over other containers – ceramic or stone vessels, woven baskets, bladder or hide sacs, shells or even large nuts – became apparent.

Air, insects and rodents were and are common enemies of the safe storage of many food staples. Wooden barrels provide protection from all of these: wine can stay bunged in barrels for several months without oxidation; water does not leak from a properly made barrel; and saltwater, insects and rats cannot enter to spoil flour, grain or meat. Some whiskies are aged for up to twelve years in the same barrel.

The ability of barrels to store salted meat safely created the need for even more barrels. Salted beef and pork, the staples of naval and merchant-ship diets during the age of sail, needed fresh water in which to soak prior to cooking and eating. Thus ships carried barrels of fresh water both for drinking and for preparing food for eating. The maintenance and repair of the barrels would have been the responsibility of the ship's cooper, a name still used to describe the craftsmen who build and care for barrels, as well as a now-common English surname. And from cooper, the word 'cooperage' evolved as the place where the barrels are made and the generic name for the round, wooden containers.

Security, mobility, adaptability, cost and ease of fabrication were all advantages that food producers and merchants sought when considering the containers for their products. The wooden barrel, whether tight or slack, filled these requirements for many historic, and a small number of present-day, commodities.

Yet barrels were not a primary product; they were the packaging. And, by extension, the demand for barrels was subject to the winds of a constantly shifting economy which bought and sold greater or lesser amounts of wine, or herring, or oil. The coopers who fabricated the barrels constituted an adjunct part of the equation. Their livelihood was determined by which container the merchants considered best for their commodity, how well that commodity sold and what regional and national politics were currently in play. Compounding these factors were the periodic episodes of war, famine, plague and specific kinds of weather – all issues continually impacting demand for the barrel and its usage.

## Making the Barrels

The evolution of wooden barrels interweaves constantly with that of wooden boats – those built with timbers and planks. The materials and tools needed to build each are similar, as are many of the woodworking techniques. The ability to split, saw and shape hard wood, such as oak, was required. This meant durable tools, capable of holding an edge and with the ability to be used with some precision. Bending wood to form complex shapes was integral to both industries. And the ultimate requirement to be an effective barrier for liquid – keeping it in or out – is identical for boat or barrel.

Many of the woods used to fabricate both boats and barrels – oak, cedar, chestnut, cypress, elm, fir, kauri, pine, spruce and walnut – have also been similar. Other woods, selectively used in various aspects of boat building, such as acacia, willow, alder, redwood, cherry, eucalyptus, white birch, poplar and linden, were also used for specialty barrels and vats.[8] It has been documented that possibly as early as 350 BC, the northern European Celtic tribes, with access in their part of the world to abundant amounts of these woods plus their ability to smelt iron and weld steel to the edges of iron tools, were crafting both boats and barrels from these timbers.[9]

Metal screws and fittings in boats were available, though not common until the sixteenth and seventeenth centuries. Similarly, the metal hoops used to hold the barrel slats, or staves, together were not in regular use until the end of the eighteenth century. Throughout most of their history, barrels have been held together by strips of flexible wood, such as chestnut or hazelnut saplings or willow boughs. These bands would have been curved into a circle, with the two ends then fastened by connecting hooked notches and that connection bound tight with willow caning: simple and ingenious. By using many wooden hoops on a barrel – historic paintings and illustrations often show sixteen to eighteen in use – a strength equal to today's use of six metal hoops was achieved.

Barrel-making processes have changed little over the thousands of years during which barrels have been in use, and barrel staves are a case in point. The staves are the long pieces of wood that make up the sides of a barrel. The staves are extracted out of a log bolt – a short piece of the log that is cut just slightly longer than the desired stave length in order to allow for shrinkage and waste – as straight pieces of wood, just as they were when the Celts started to make barrels, and are then heated. Heating allows them to be curved into the shape required for the barrel. These processes are in common with those used for boat building.

The main use for wooden barrels today is for the ageing of wine, including fortified wines like sherry and port, and spirits, such as Cognac, Armagnac, whiskey, Scotch whisky, bourbon and rum. Today's wine and spirit barrels are made from both European and American white oaks, while for whiskey and bourbon, American white oak is the primary wood.

Producing the staves and heading from oak differs depending on whether European or American oaks are being used. This is primarily due to the oaks' intercellular structure, which affects the potential

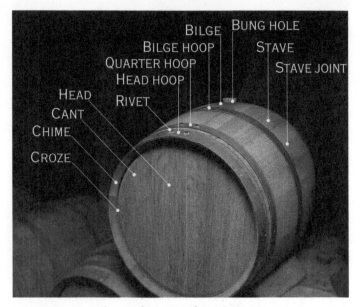

The terms for the parts of a wooden barrel.

for leakage in the final pieces. The cellular structure in American white oak allows for the timber to be sawn in order to produce the staves and heading, while to extract the staves and heading from European oak, the wood needs to be split in order to minimize leakage.

## Different Uses, Different Sizes

Throughout the more than two millennia during which wooden barrels, and other wooden containers – from small buckets to huge tanks – have been constructed, the majority have been manufactured for specific commodities. The quality, quantity and weight of the commodity usually dictated the size and shape of that container, but occasionally the resources, or lack thereof, of a cooper also played a critical role. Barrels come in hundreds of sizes, most with their own unique names, often related to the product they contain – cracker barrel, nail keg, beer barrel and so on. To simplify the naming system used in this book, small barrels with capacities of up to 90 litres (25 gallons) will be called kegs, and the commonly seen wine and whiskey barrels with capacities of 90–225 litres (25–60 gallons) will be termed barrels. Casks, butts and hogsheads, and the French term 'puncheon', are all names for containers with capacities from 225 litres (60 gallons) up to about 1,100 litres (340 gallons).[10] The use of the terms 'vats' and 'tanks' is less precise, both being used for containers of 560 litre (150 gallons) upwards, while 'tank' is generally only used for containers of thousands of litres or gallons. Vats usually do not have a cover, while tanks are normally fully enclosed containers.

Wooden barrels are not just highly practical as containers; they contribute some wonderfully complex flavours and aromas to wines and spirits and are still highly important for ageing the world's finest wines and whiskies. Cooperage, or the world of barrels, like so many forgotten crafts, is a mystery to many people today. Extremely common not that many years ago, we now see barrels pictured primarily in nostalgic ads for wine or whiskey or in our garden centres, cut in half and intended to be used as planters. Today, few non-specialists know how barrels are made, so to provide a fundamental understanding of barrel construction, it is helpful to begin by considering their basic design and usages.

If you had to design a barrel, would you make it round or rectangular? Would it look like the typical round, bulbous barrel seen in a winery advertisement or have some other shape? If the crucial criterion is to maximize space, you would most likely design a

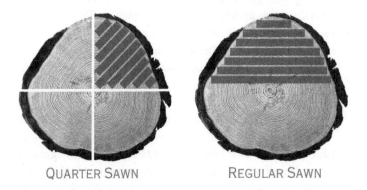

QUARTER SAWN          REGULAR SAWN

Examples of quarter-sawn and regular (also known as plain or flat) sawn timber.

rectangular barrel with six flat sides. In fact, square barrels have been attempted over the ages, and in 1969, one George Morris was issued a United States patent for a design. However, a square barrel has rarely, if ever, been utilized because the hoop-like fasteners are complicated, and it is difficult to move. And making a rectangular barrel so that it contains a liquid adequately is extremely complicated. Liquid containment is far easier with the round barrel: as the hoops are pressed tighter on the barrel, they are essentially squeezing all the staves and the two flat-surfaced heads – in essence only three surfaces – at the same time.

You might also suggest that the staves be secured and tightened with a band (hoop) utilizing a screw or bolt connector. This is all well and good for today's barrel, but would not have solved some of the problems of those people, over 2,000 years ago, who were attempting to construct the first barrels. For securing the bands, they couldn't go to the local hardware store for screws or bolts. Instead, for the prototype barrel, which was basically a pail or bucket, their need for a simple closure was resolved by tapering the sides and then applying smaller and smaller bands which simply got tighter as they were pressed on. And if the bands were of wood they would have been lashed together, or, if of iron, they would have been riveted – both far simpler connectors than screws or bolts.

To meet the desire to make the buckets bigger and to be able to enclose them (for example placing two buckets together, open end to open end, in order to retain the tapered sides), it becomes obvious that the staves need to be bent. The problem of how to do this was solved by heating the staves, as several species of wood, in the correct orientation, can be bent without breaking or cracking. The staves span the length of the body of the barrel, bulging outward at the equator, or bilge, of the barrel. Thus the barrel is really a series of parallel arches – the curved staves – assembled together. Structurally these bowed staves, or arches, add significant strength to the barrel.

Our barrel now is round, with curved sides and hoops or bands that do not need a special fastening device but can be easily tightened just by pushing them on a bit more. It also turns out that the curved sides provided another exceptional advantage – making barrels easy to roll. This benefit was essential for moving barrels in an age prior

Demonstrating the skill of rolling two barrels at the 2012 Bordeaux Wine Festival.

to that of today's cranes, forklifts, pallet jacks or other mechanized equipment. Rolling the barrels on the apex of the curve, the bilge, was relatively easy, even with a heavily loaded barrel. Plus, with such a small pivot point, the barrels could be turned and manoeuvred while rolling.

The bent staves also have a little give, or flexibility, opening slightly when the hoops are removed to allow the heads (ends) of the barrel to be inserted and removed. This is particularly helpful when loading or unloading solid commodities, such as salt, flour, meat, apples or hides, from the barrel. The heads also provide a stable platform upon which the barrel can rest to keep it stationary in a warehouse or ship. In barrels that are to contain liquids, a bunghole is drilled in a wide stave at the bilge, or in the head, and is used for filling and emptying the liquids.

The barrel's story has origins which start well before the Celts began crafting what we now know as a barrel, and continue right up to ageing wine in today's most elegant wineries or chateaux, bourbon in a Kentucky warehouse or Scotch in a Highland distillery. The narrative includes the work coopers perform and how they blend science, art and skill to make a watertight, durable wooden container; the rise of the cooperages and the key roles of barrels aboard ships; how this versatile container is used in the production of some of the world's most desirable beverages; and the innumerable uses for this amazing, and enduring, piece of craftsmanship.

# TWO

# Evolution: From Buckets
# to Barrels

The earliest type of barrel probably was the one consisting
of a log hollowed out and the end covered with skins.

Franklin E. Coyne, *The Development of the Cooperage Industry
in the United States, 1620–1940*

 A graph profiling the use of barrels reveals the slow, almost
imperceptible ascent up the left-hand side of the graph,
with a dramatically steep descent on the right. A line
showing the development of wooden boats would follow roughly
the same pattern. Several millennia before the present, they both
start as rudimentary items – tubs and buckets in small village homes
and nomadic tents and the small sailing boats and barges that
could be found along the shores of the Mediterranean and the rivers
of Asia.

For the barrels, the graph would start more than 2,000 years ago.
At the far left, the rising grade of the line is very gradual: just a few
primitive barrels in use by the various Celtic tribes of northern Europe.
It then starts to ascend as the Roman civilization expanded, incorporating the Celtic culture and technology of barrel making for their
trade in wine and other commodities. The rise continues, albeit a bit
more steeply, for the next 1,000 years –as the population of Europe
enlarges, as more and more products are stored and shipped in
wooden barrels and as trade expands outside the Mediterranean and
European regions.

From roughly AD 1000 to 1400, there is a dramatic rise in the line.
The use of barrels increased for domestic trade and for the transportation of goods and supplies as ships started to sail forth on extended
voyages that required watertight, relatively secure containers. The
heyday of barrels, and of wooden sailing ships for that matter, was
from the fourteenth to the early twentieth century, when a huge

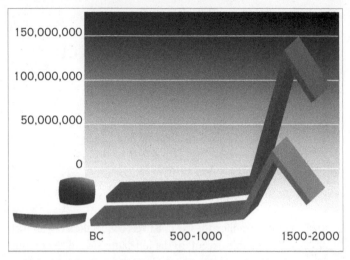

An approximate timeline of the development of wooden barrels, with estimates of total numbers of barrels and wooden boats in use throughout the ages.

number of commodities, both liquid and dry solids, were transported, aged and stored in barrels.

Then, within a few decades, after taking over 2,000 years to reach the apogee of the graphic curve, where millions of barrels are in use worldwide and barrels are one of the most common large, bulk containers, the bottom of the barrel market, figuratively speaking, drops out. The graph line makes a rapid descent. By the late nineteenth and early twentieth century, barrels are rapidly replaced by other packaging materials – cotton sacks, paper bags, cardboard boxes, glass jars and bottles, metal cans and kegs and finally plastic, in its multiple forms. The worldwide demand for barrels falls from tens of millions per year to under two million per year. Although far fewer in number, wooden boats also experienced a similar decline in popularity with the advancements in steel, aluminium and fibreglass technology.

## Barrel Development: Prehistory

The earliest recorded evidence of wooden pails or buckets, the forerunners of barrels, has been noted by Diana Twede in her excellent article about the barrel as a container. She related that small wooden buckets and tubs have been found in Egyptian tombs.[1] In a tomb dating from 2690 BC, a wooden bucket was discovered which was used to measure corn, and on the wall of another tomb, a mural dated to 1900 BC depicts a wooden tub used for holding grapes during the harvest. Alongside this small vat is a man who may have been the cooper who constructed it.[2]

Those early buckets were crafted from soft woods, such as palm, pine or poplar.[3] A hole in the centre of a short section of a log might have been carved with tools made from stone – those which could hold an edge such as flint, argillite or obsidian. Later, tools made from relatively soft copper, or harder bronze, could have been utilized to carve away the interior wood. An alternative to carving a bucket would have been to burn out the inside. Any or all of these techniques might have been concurrent and, often, the skills and techniques learned in one culture, or in one generation, are lost in another, only to be relearned or rediscovered later.[4]

Once one-piece buckets were made, a subsequent evolutionary step would have been to repair a bucket which had been cracked. It would have been tough to do if the sides were cylindrical: prior to screw-type fasteners, there were no simple tightening mechanisms. However, if the bucket had a taper (the form we see today in metal buckets and toy plastic beach pails), a hoop of sapling or fibre rope with a diameter just larger than the bottom of the bucket could be placed over that end. Tightening as it was pushed on, it would close a crack or gap. So, the concept of a cylinder with a taper would have been a critical milestone in this developmental process. Another minor, but important, technique: at some point someone would have realized that if the sapling hoop was wet, it would stretch. Then, when it was snug on the bucket, it would shrink as it dried, becoming extremely tight.

Design-wise, a reverse taper, one with the bucket's smaller end at the top, would have made better sense in terms of the hoops – if the bucket dried and the hoops loosened, gravity would pull them down to make them tighter. This shape is sometimes employed in large vats and tanks for wine or water, and, until recently, was still prevalent in Russian- and Canadian-style pails.[5] But for a pail or

Illustration from a manuscript, *c.* 1270, of the Martyrology of the 9th-century Benedictine monk Usuard of Saint-Germain-des-Prés, showing a wooden tank in use during the grape harvest.

bucket, a smaller diameter at the top is not as practical as one with a wider opening, both for filling and for emptying.

At some further point, if the bucket broke again, perhaps someone thought of replacing the broken section. Cutting and planing a smooth joint on the replacement piece with stone tools would have been difficult. It is likely that this stage in the development did not occur until the development of copper or bronze tools.

This brings us to the concept of the buckets and tubs found and depicted in the Egyptian tombs – that of placing several pieces of wood together to actually form a bucket. Again, all the pieces could be held together with saplings or fibre rope, and the bottom would have provided a circular form for the tapered and edge-angled side pieces around which they would be assembled. Tool-wise, a saw or adze would have been required to shape each piece of wood roughly and some type of plane or draw knife needed to smooth the joining edges. Also required of the crafting process would have been to cut a groove or notch along the lower flat edge of each stave in order to attach them to the bottom section. This groove could have been made with a chisel or a more sophisticated plane.

An often-overlooked fact when making buckets or barrels is the need for sufficiently dry wood to avoid any shrinkage once the container is assembled, especially for barrels holding liquids. Depending upon the type of wood and the ambient climate, it generally takes two years in the open air for wood to become thoroughly dry. In arid regions, caution is also necessary to ensure that the wood does not dry too quickly and crack prior to being put to use. This prevention process can still be seen today at lumber mills where logs are soaked in a pond, or watered with a sprinkler system, to prevent the rapid drying while awaiting cutting. Another current example: once the wood is cut into dimensional pieces, modern woodworking operations often employ dry kilns to speed up the drying process. Taking weeks or months instead of years, the wood in the kilns is dried under extremely controlled conditions utilizing incremental adjustments in the air temperature, humidity and circulation. For yesteryear's coopers, it was important to set the wood out well in advance of its anticipated use, as it is for today's wine-barrel cooperages wanting solely air-dried wood.

The evidence suggests that somewhere between c. 1000 and c. 500 BC, people from the Celtic culture of transalpine Europe advanced the multi-piece bucket idea enough to actually develop and make a barrel. Given the extent of their technologies, it appears that they were able to assemble a number of pieces of shaped wood, bend them over a fire, fit wooden discs on the ends and enclose the resultant container with hooping material – either saplings or metal. This was not a simple idea. It must have seen multiple attempts, with many variations in the designs and wood materials. The historic data also suggests a surprising amount of inter-tribal trade and, by extension, the exchange of ideas. This communication would have aided the Celtic craftsmen, as they would have benefited from each other's slightly different version, the whole process leading up to the eventual development of what we now know as the barrel.

The main new concept was one of basically placing two buckets end to end. And while the concept of bending wood was, at that stage, probably not new, certainly it took someone's ingenious imagination to envision the curved staves as the connection of placing two buckets together, open end to open end. Details such as how to bend the wood without breaking it, getting the liquids in and out of the barrel and how to move a barrel – concepts which we consider so simple that they do not even register in our consciousness – at some point must have been part of the painstaking developmental process.

Additionally, while stone, copper and bronze tools may have been adequate for soft woods such as pine, the concept of a complete barrel may not have advanced until refinements in woodworking tools occurred concurrently, especially for slightly harder woods such as fir or tougher woods such as oak. The techniques of smelting iron, and making harder steel, were slowly making their way northwest via the smiths in the Celtic tribes. From iron, these craftsmen were developing the tools which were required to cut and shape the wooden pieces needed to assemble the wooden wheels for their chariots and wagons, as well as the timbers for their boats, buildings and furniture. To shape each of these items accurately it would have been necessary to smooth the surfaces to be joined. This was equally true for both stave edges and boat planking, in order to make them watertight. Early Egyptians, Assyrians, Babylonians and Greeks had some of these tools, but the woods they worked were softer than the northern hardwoods. It probably wasn't until the apogee of Celtic culture within the European forests, from c. 1300 BC to c. AD 200,[6] which brought stronger iron tools suited to working the harder northern tree species, that these refinements in both boats and barrels could be achieved.

## Amphorae: Former Packaging Containers

Prior to and during the period of barrel development, another bulk-packaging container – the amphora – was an extremely common feature of the trade between Mediterranean communities. It is worthwhile here to examine the origin and use of the amphora as a comparison to the barrel. This overview can provide insight into the issues of how an entirely new container and packaging concept – the barrel – could enter into the mainstream trade and transportation systems of the first millennium BC.

Evidence of the first simple pottery – from shards and broken pots – has been dated to c. 14,000 BC, possibly as early as 18,000 BC. Advance a couple of thousand years, and the amphora, Ian Morris suggests, was developed by the Chinese in around 4800 BC, with containers of a similar shape being found later in the West, starting c. 3500 BC in Syria.[7] Although now all but phased out, amphorae have been in use for some 6,000 years.

Archaeologists, digging through ancient town dumps and burial sites, have discovered that the early hunter-gatherer societies began to make and use pottery for storage and household implements once they settled into permanent camps and villages.[8] At first it was simple bowls and pots; later, as the towns and villages grew, ceramics developed stylistically and for different uses. Contrasted with the nomadic life of hunter-gathering, a more stationary existence required longer-term storage of foods due to the decreasing ability to move around in order to gather provisions.[9] This need for increased storage capacity necessitated larger storage containers and different storage methods. Additionally, the erratic extremes of the food supply encouraged trade in order to even out the abundance or scarcity between one community and another. Around the Mediterranean, the Greeks, and then the Romans, used amphorae to store and transport wine, olive oil, olives and grapes, fish, grain and other commodities such as dates and nuts.[10]

Amphorae come in many shapes, but the long, thin shape, with the pointed bottom, narrow opening and two handles, is the one we commonly associate with trade. It is a shape that evolved partly through trial and error, partly through necessity, probably with a bit of ceramic artistry thrown in for good measure. Certainly, some sort of style with a small closure was necessary for air-sensitive items such as wine, olive oil, fish oil and small fish delicately preserved in olive oil or brine. And a need for durable containers to transport

Styles of amphorae used to transport olive oil and wine.

these liquids, as well as other commodities, from the often far-flung producers to the expanding towns and cities around the Mediterranean was an additional stimulus. Part of the design equation also included utilizing existing sea and river transportation systems, which provided a far more efficient method for the long-distance movement of bulk products before the days of improved Roman roads. The use of the amphora for grain and grape transport does not sound particularly practical, but then perhaps it was convenient: amphorae were available within the transport system for other commodities. We will see barrels similarly being utilized and recycled for multiple products.

Twede notes that the egg-like form of the amphora is the strongest shape that can be made from clay; the pointed toe absorbed shocks and there is evidence that potters reinforced the area above it, which was the most vulnerable to breakage.[11] She observed that amphorae nested well, interlocking with each other to form a reasonably tight and secure load within the flat-bottomed ships which plied the Mediterranean Sea. Additionally, amphorae could be placed upright in the mud flats and sand beaches abutting the rivers and seas while awaiting transport or after off-loading on arrival. Another reason for the pointed bottom was to conserve weight; a large, flat base would have added significantly to the amount of materials used for manufacturing, and ultimately to the weight of the amphora.[12]

Like barrels later on, the amphora was the primary container for the expanding regional and global trade. Around the Mediterranean, many amphorae contained wine traded from Italy and Greece, and in Rome, amphorae have been found to have transported olive oil from Spain and grain from Spain, Tunisia and Libya. Roman-style amphorae have been found as far afield as Britain and India.[13]

In some historical sites, such as Tolosa, Spain, upriver from San Sebastián, and Piscul Crăşani, Romania, adjacent to the Ialomiţa River, a tributary of the Danube near Bucharest, huge piles of broken and smashed amphorae have been discovered.[14] Archaeologists surmise that wine, which was in transit via the rivers to inland regions, was transferred from amphorae into goat skins or small kegs for the remainder of the trip, carried by pack animal. In the early days of the Roman Empire, Italy was even shipping wine in amphorae to the Celts in France and Germany! John Davis notes that 'It is estimated that forty million *amphorae* were imported into Gaul in the century after 150 BC.' And he continues: 'Around Toulouse, there are fields so cluttered with *amphora* shards that, even today, ploughing presents problems.'[15]

The huge historic dump at Rome's Monte Testaccio contains an estimated 53 million amphorae. Monte Testaccio is located in Regione XIII Aventinus, near Rome's ancient port. The amphorae were dumped at this site starting in the first century BC. Evidence indicates that its use tailed off in the third century AD, coinciding with the decline of the Roman Empire and as barrels became common.[16]

The majority of amphorae in Monte Testaccio came from the Spanish olive oil and grain shipments, used to feed first- and second-century AD Rome's voracious appetite for goods and commodities. The oil was used as a fuel for lamps, with foods and as an alternative to soap – after a bath in olive oil they just scraped it off.[17] Within the huge pile at Monte Testaccio, the primary style of amphorae is the Dressel 20, which was made in workshops in the southwest Spanish regions around Seville and Cadiz. Other amphorae in the mountainous dump came from Libya and Tunisia. The capacity of most of these amphorae was around 70 l, and when filled, they each weighed about 37 kg.[18] With an estimated 130,000 amphorae used each year, this enormous pile was established by the Roman authorities as a sanitary measure to dispose of the amphorae, and the Dressel 20 type in particular. Once the amphorae were received at this ancient Roman port, and the oil transferred to other containers for distribution within Rome itself, any leftover oil in amphorae would turn rancid, while morsels of grain would promote rats and mice. Therefore, rather than just leaving them around to stink and clutter up the city, this dump provided a controlled method of disposing of these empty containers.[19]

The Dressel 20 was apparently not suited to recycling for other products, in part because the oil would impact dry products. Strangely, due to either the design or the type of clay, it was not effective to break them into tiny bits to be used in concrete, as was the case for some of the other types of pottery, or to reuse them as water piping, as were other amphorae styles. Again, we will see similar recycling efforts made with barrels, as well as some comparable difficulties.

In other eras and regions amphorae were recycled. Within Egypt, for example, after Greek and Phoenician wine was received and dispensed, the amphorae were subsequently filled with water, which was sent to desert tracts in Syria.[20] And there is some evidence that those amphorae, filled with Syrian dates and nuts, were sent back to Greece.

The early amphora types were made by stacking coils of clay on top of one another. However, as demand for them increased, especially during the Roman era, manufacturing turned to making them on potters' wheels or fabricating sections within jigs and forms and then sticking the pieces together with a ceramic slip.[21] To make the numbers estimated in the Monte Testaccio dump, huge quarries for the clay and workshops of up to ten men at numerous sites near the seaports would have been required. This is quite different from the fabrication of barrels in the latter Roman period, and even well into the Middle Ages, where barrel cooperages of only one or two men were the norm. For barrels, it was only once the demand for salted fish and whale oil was well established, starting in about the fourteenth century, and, later, when using the containers to hold petroleum commenced, that cooperages evolved and began to employ significant numbers of workers.

Along with the size of Rome's amphora landfill, the number of amphorae found in Mediterranean shipwrecks provides further evidence that amphorae were developed and refined for trade. Roman merchant ships used amphorae for most of the cargo, including wine, grain, olive oil and fish sauce (garum).[22] Leslie and Roy Adkins state that most of these ships were small, carrying loads of 150 metric tons. However, some wrecks have been found to have been transporting far greater amounts – the Adkins mention loads of up to 5,000 amphorae. A shipment of that quantity would have weighed at least 181 metric tons, with a potential capacity of 3,406 hectolitres of wine – certainly not an insignificant amount in the early days of commercial shipping. With such a large product consignment, it is easy to understand how some Roman traders became wealthy.

Twede contends that the amphora was the first 'consumer packaging'.[23] The conspicuous and unique size, style and inscriptions of these vessels provided the traders, retailers and ultimately the end consumers with the origin, type and grade of the product they were purchasing. And in Rome, at least, there was pressure for standardization, which provided a degree of consumer confidence in the quantity and quality of the product. Similar issues were raised with barrels as they came into mainstream commerce. The unique characteristics of each type of amphora even assisted boatmen, dock and warehouse workers, many of whom were illiterate, in determining which amphorae to move or stack.[24]

Wooden barrels have had similar identification systems. It was normal for coopers to carve their mark, or sign their name, on each item they made. However, the nature of a barrel's contents would be known to the tradesmen and shippers by its size and style, which was often standardized within a particular trading system for a specific commodity. Individual shipments or stores of barrels were numbered. However, no attempt was made to number or barcode every barrel in the world, as with today's metal shipping containers.

Throughout history, Morris notes: 'Change is caused by lazy, greedy, frightened people looking for an easier, more profitable and safer way to do things. And they rarely know what they are doing.'[25] The amphora fulfilled a need, and that need was met by the materials and resources of the day. Despite their attributes, amphorae were phased out of major use in the Mediterranean after about AD 500.[26] Twede does note that amphorae are still used for water and wine in some rural areas of the Mediterranean.[27] However, after AD 500, perhaps the times and needs had changed to a point where the wooden barrel and other containers could take over in many regions, in much the same way as barrels were displaced by other containers and materials from the 1900s onwards.

# THREE

# Celts: A Nexus of Skills and Technology

The hand tools which we still use today were all anticipated
by Celtic craftsmen at least by the first century BC.

Peter Berresford Ellis, *The Ancient World of the Celts*

 Somewhat like the hidden clues in a good detective novel, the circumstantial evidence, in the form of historical documents that point towards the development of the wooden barrel within the Celtic culture, overshadows the physical relics. Yet this documentary evidence does not come from the Celts, who at the time had no written language; it is derived largely from their contemporary Roman historians. The few barrels, pails and buckets which have been uncovered indicate that cooperage was relatively common and advanced. However, researchers are still puzzling over the details of its developmental progression and its role within the everyday life of Celtic tribes and communities.

For the documentary evidence, one significant anecdote comes from the multi-volume *Natural History* written by the Roman historian Pliny the Elder, who lived from AD 23 to 79. His far-reaching work describes the 'best practices' in a range of subjects. The era's winemaking techniques and technology are discussed extensively, as are numerous details on such diverse topics as agriculture, animal husbandry, gardening, apiary, carpentry and human medicinal and herbal health practices. In his section on 'Wine Vessels and Wine Cellars', which discusses the contemporary winemaking procedures used in northern Italy, he wrote: 'In the vicinity of the Alps, they put their wines in wooden vessels hooped around.'[1] The English translators of his books, writing around 1855, surmise that he was referring to wooden barrels, stating in a footnote that the wooden vessels would be 'casks, in fact, similar to those used in France at the present day'.[2]

The few barrel artefacts which have been uncovered, dating from several centuries prior to when Pliny was writing, plus the fact that there were Celtic tribes living in the Alpine regions to the north of Italy, would indicate that it was those Celtic people, and their wooden barrels, to whom he was referring.

The physical evidence for Celtic barrel development, though it is limited, includes several Celtic-era barrels and other wooden artefacts which have been preserved within peat soils, in the very cold Alpine lake waters and in some Celtic tombs. Ancient metal-hooped, multi-pieced buckets and pitchers have been found.[3] Barrels made of Alpine silver fir (*Abies alba*) have been excavated in the ancient Celtic city of Manching in south-central Germany.[4] Other barrels have been found used as linings for water wells in Roman-era Britain[5] – believed to have been utilized for this after transporting wine to Britain, despite a lack of wine residue in the staves. British cooper Ken Kilby has noted an advanced type of bucket artefact found in the Lake Village excavation near Glastonbury in the UK. Dating from the late Iron Age, prior to the Roman occupation of Celtic Britain,[6] this was a tub 17.5 centimetres in diameter by 16 centimetres tall, formed with twelve staves. Several of the staves were longer than the others and were intended as legs to raise it above the ground. The staves had been dowelled together – not an uncommon method in bucket assembly. What is evident from this find is the high calibre of craftsmanship and the advanced woodworking techniques, both of which are indicators of the ability to make larger barrels.

To understand barrel development, it is useful to examine Celtic culture in the millennium before Christ. The ancient Celtic community of Heuneburg in south-central Germany sits on a hill overlooking the Danube River, about 80 km south of Stuttgart. As a recent visitor to this site, when standing atop a recently reconstructed wall, I had a clear view of the Danube and its surrounding floodplains – no doubt a similar view to that seen by Celtic sentries over 2,000 years ago. Then, with some of the forests cleared of timber for construction of dwellings, cooperage, furniture and firewood, the plains would have been a patchwork of cow pastures and cultivated land. Travellers using the river for transport, or passing overland on foot, horseback or with a cart through the Danube valley, would have been observed. And probably much traffic passed Heuneburg as it was centrally located within the greater Celtic transalpine community, about 200 km upriver from the city of Manching, another important historic Celtic site. Today, Heuneburg serves as an educational centre for Celtic culture, with some restoration of the fortress wall and several typical buildings.

At its pinnacle, Heuneburg was a 'princely residence', a hill town where the Celtic elite and nobility lived.[7] It was also an *oppida*, meaning it was fortified and surrounded by a large block wall.[8] Both the residence and block wall are critical clues towards understanding the period's cultural climate and how this might have enabled the Celts to conceptualize and construct barrels of the design which we know today.

Archaeological evidence indicates that a community had lived on the hilltop for hundreds of years. By *c.* 600 BC, the Celtic culture was, in P. B. Ellis's words, 'a civilization which had developed from its Indo-European roots around the headwaters of the Rhine,

Celtic-era barrels, found lining water wells
in the Roman-era town of Silchester, England.

the Rhône and the Danube [and] suddenly erupted in all directions through Europe.'[9] This cultural explosion was due largely to Celts' highly developed knowledge of ironwork, enabling them to create hard-wearing tools, durable weapons and the metal fittings for speedy chariots, all of which seemed to have spurred them into becoming 'a powerful and irresistible force'.[10]

Information about the Celtic nobility of Heuneburg, as with many of the Celtic *oppida* sites throughout Europe, is gleaned from the excavations of the nobles' graves and from other ancient Celtic town sites that have provided numerous iron and ceramic artefacts. Much has also been learned about the Celts from other Greek and Roman historians and writers, including Aristotle, Plato, Diodorus Siculus and Julius Caesar, although their commentaries must be read in light of each author's particular cultural bias and motives.[11] To understand the development of the barrel, what becomes clear in examining the types of artefacts, and from reading the histories, are three somewhat interrelated, circumstantial factors: the Celts loved wine and beer;[12] they were highly skilled at ironwork, which enabled the creation of precision tools required for working the harder northern timbers; and, despite their constant warring with their neighbours, they had numerous interactions with those same neighbours via the wide-ranging trade of goods and ideas.

Because the Celts had no written language, descriptions of their daily lives, and specifically that of the coopers, are only found in accounts written by the Romans. We do know that the Celtic tribes had a well-developed tradesman class.[13] Their farmers produced surplus crops that lasted long enough to allow craftsmen the time to make products and offer services. As their settlements grew into villages and towns, these craftsmen – the blacksmith, woodsman, miner, cooper, carpenter, mason, forester, wheel maker, cart maker, miller, baker, and so on – contributed to each other's skills. For instance, the blacksmith could forge tools for the cooper, while the cooper could make a water barrel for the blacksmith to cool the wrought-iron pieces he worked. These interactive contributions led to the development of many items, including barrels.

## A Love of Wine

The archaeologist Barry Cunliffe notes that a significant number of the artefacts found in the graves of Celtic nobility were related to wine drinking: goblets, pitchers and *kraters* (bowls for mixing the wine with water).[14] This indicates that eating and drinking were an important part of the social life of the Celtic elite.

The Celts' love of wine is supported by the discovery of wine-related equipment from the era and by writings from some contemporary historians. Plato wrote of the Celts' penchant for wine: 'They craved it.'[15] Another Greek historian, Diodorus Siculus, writing between 60 and 30 BC, noted: 'They sate themselves with unmixed [without added water] wine. Their desire makes them drink it greedily and when they become drunk they fall into a stupor or into a manic disposition.'[16]

With this craving for the elixir of the grape, it is no wonder that Roman merchants happily shipped boatloads of wine-filled amphorae from the trading centre at Massalia (now Marseilles) up the Rhône and spread their wares out into what is now France and Germany. Other traders routed wine to the east into Hungary and up the Danube and to the west into Britain. Payment for all this wine came back as slaves, furs, salt, gold, tin for making bronze, cattle and hides and iron ingots and blooms, the rough hunks of basic iron weighing between 5 and 10 kg that, when reheated and hammered, could produce all manner of weapons and tools.[17] The Celts' love of wine, and the trade which brought it, is further evidenced, as noted in the section on amphorae, by those thousands of crushed and broken amphorae found in historic sites throughout the northern European Celtic homelands.[18]

By today's standards, Roman wine, mixed or unmixed, was probably pretty rough stuff. Despite the great detail Pliny the Elder provided in his *Natural History* about growing grapes and making wine, there was no knowledge of the microorganisms which interact with wine, for better or worse. But the vintners must have told Pliny that some sort of sanitation was required. He therefore suggested the following for cleaning wine-storage vessels:

An engraving from Diderot's *Encyclopédie* depicting wooden barrels, buckets, pails and tubs in use in the 18th century. Their prototypes were developed during the Celtic era.

to pitch [empty] them immediately after the rising of the
Dog-star, and then to wash them either with sea or salt
water, after which they should be sprinkled with the ashes
of tree-shoots or else with potters' earth; they ought then
to be cleaned out, and perfumed with myrrh . . .[19]

Pliny's recommendation for cleaning with salt water and ashes
would have constituted a good start at removing harmful bacteria.
In Italy the dog star, Sirius, rose in the night sky near the summer
solstice and the beginning of the summer heat.[20] It may also have
been an indicator of when to empty the wine containers, and perhaps
when to ship the wine to market. Unlike at today's wineries, winemaking in those early days was largely an outdoor activity, with the
crushing and fermentation all done in the open, or maybe at best
under a roof. Emptying the wine out of big ceramic pots or tanks
and placing it in smaller ones, or barrels, which could then be kept
in the consumer's cellars, and therefore out of the Mediterranean
summer heat, made good sense. While the suggested practices for
sanitary winemaking may be questionable in some ways, the underlining rationale for Pliny's recommendations is sound.

Coupled with the minimal sanitation during winemaking and
almost no climate control during the transportation of wine, grape
growers during this period also had little control in their vineyards:
pruning techniques and sulphuring were still in their infancy, and
frost protection would have been rudimentary at best. Providing
a consistently palatable product must have been a real challenge for
the Roman vintners. Wines with higher alcohol, perhaps 14 per
cent or more, stood a better chance of remaining somewhat drink-
able by the time they reached the end-user. To tone down the highly
alcoholic wines, and perhaps to stretch such a prized commodity,
the Celts often diluted their wines with water prior to consumption.

Besides the availability of amphorae within the trading system,
there was another important reason to utilize them to ship wine.
The wine growers around the Mediterranean, and the Celts receiving
the wine-filled amphorae, would have known that wine spoils when
exposed to too much oxygen, though it is unlikely that anyone at
that time understood the science behind the causes of oxidation.

But they did know enough to seal the wine from oxygen, and with its small opening, the amphora was a suitable container. The small aperture in wooden barrels, the bunghole, was designed for the same reason.

The wine paraphernalia found by archaeologists in the Celtic tombs consisted primarily of vessels for consuming wine, not making it. Within the graves of noble men and women, such as those at and near Heuneburg, the types of artefacts discovered were the bronze, wide-mouth *krater* basins for mixing wine with water; jugs for pouring wine; and the ceramic Attic wine goblets, imported from Greece via the trade routes.[21] While the Celts were basically farmers,[22] they imported wine and didn't grow their own grapes. Although the Ice Age had retreated thousands of years before the Celtic people blossomed in Europe, anthropologist Brian Fagan notes that the climate of Europe was still very much unsettled. The Celts survived by 'their highly flexible farming strategies and cattle-herding practices [which] were well suited to an uncertain climate'.[23] Despite the wild grapes growing in northern Europe, apparently neither the cold-hardy wine grape varieties, nor the understanding of how to manage grapevines through the harsh, north-of-the-Alps winters, had yet to be fully developed. It was only later, once the Romans had control of the regions, that wine grapevines were introduced into Bordeaux, Burgundy, the Mosel, the Rhineland and Britain.[24]

Also related to barrel development was the Germanic Celtic tribes' love of beer[25] – the drink of the general public. One Roman-era brewery was unearthed in Regensburg, not far from Heuneburg, and another near Alzey, in what is now Germany's largest wine-growing region, in the state of Rhineland-Palatinate. Among the artefacts found were wooden vats and tanks to make the mash and ferment the beer.[26] While beer barrels, the smallish, 70-litre style used to transport beer in the eighteenth and early nineteenth centuries, may not have been in general use, the larger vessels incorporating several of the cooperage craft technologies are evident in these relics.

## Iron Technology

Another key item in charting the development of barrels has been the discovery of iron tools in the Celtic graves and excavated sites. According to Morris, iron production in the eastern Mediterranean was known prior to 1200 BC, but bronze was still the most commonly used metal.[27] Around 1000 BC, in one of the many collapses of Mediterranean civilizations, the transportation of tin supplies was disrupted. Tin was a critical ingredient in the smelting of bronze; smiths added it to copper in order to make the resultant bronze harder. The loss of tin was significant for the smiths on Cyprus, who, Morris states, had at the time the 'world's most advanced metallurgy'.[28] Without the tin to make bronze, Cypriot smiths were forced to make iron and subsequently improved the process of its production. Once smiths started to provide harder iron weapons to the armies and iron tools to the craftsmen, bronze implements took a back seat.

The advantages of iron over bronze and copper were evident to soldiers, farmers, coopers and other tradesmen: iron was stronger and more durable, especially if the tool or weapon had an edge of hardened steel. And tin sources were not always near those of copper for the production of bronze.[29] R. F. Tylecote, a metallurgist and historian, states that, following this shift to iron, iron production probably started in earnest around 1500 to 1000 BC in the Middle East and then spread to 'Europe, Asia, and North Africa in the following five centuries',[30] reaching the Celtic communities by 800 BC and Britain by 500 BC. By the end of the first millennium BC, the array of hand tools incorporating iron – as blades or other critical elements, available for carpenters, blacksmiths, coopers, masons and furniture makers – was similar to that found in any quality hardware store today.[31] Celtic craftsmen used axes, hammers, saws, horseshoes, locks and nails, and their smiths also incorporated iron into armour and swords for the military.[32] Additionally, iron wheel hubs, rims and other wagon fittings were available, as evidenced by iron artefacts found in the extensive excavations of Celtic tombs and Celtic village sites such as Hallstatt, Austria, and La Tène on Lake Neuchâtel, Switzerland.[33]

Another piece of the puzzle falls into place. The Celts now had iron and, with it, knowledge of how to create and weld the harder steel onto an iron block in order to strengthen the edges of weapons and tools. Numerous tasks such as cutting down trees and dressing logs, extremely difficult with bronze tools, could be now accomplished with greater speed and precision thanks to iron tools. And this would not just have applied to the softer pines, but equally to oaks and other northern European hardwood species, which could now be fully exploited. With these hard tools, the Celts' woodworking capacity must have expanded dramatically. The wheel, already in use throughout Europe prior to the Celts' emergence, was improved by Celt smiths and wheelwrights. Utilizing their iron tools, they were able to lighten wheels by replacing the solid centres with lighter, carved spokes and a wooden rim, which was a complex interconnection of multiple pieces of wood. For the chariots and wagons, they fabricated bent-wood sides and railings. The bending concept necessitated cutting wood to a specific grain orientation – yet another distinct technology that was developing. Manufacturing the pieces for the wheels required the use of adzes, saws and precision planes. With their ironwork, the Celts added iron rims to the wheels for durability, attaching them to the wooden rims with a kind of rivet.[34] This round iron rim, and the rivets used to secure it, would have been very similar to the barrel hoop and its fasteners. All these skills and tools would have facilitated making barrels.

Improved wheels, and better tools with which to work metal and wood, upgraded the carts and wagons, which in turn necessitated the construction of roads.[35] Archaeologists have recently discovered that the Celts did indeed construct all-weather, corduroy (log) roads prior to being conquered by the Romans: several have been uncovered in bogs, their ancient wooden planks protected by the acidic soils. And they also discovered that many of the Roman roads were built on top of those constructed by the Celts, which is why the Celtic roads have only recently come to light.[36] Both the advanced tools and improved roads increased regional trade.

## Trade

The last item in our puzzle of how the Celts might have developed the barrel was the incredible expansion, during that period, in the movement of goods, services and ideas. For example, Greece-sourced Attic wine-drinking goblets, found in some tombs and graves, had been transported from as far as 1,500 km away. As for the spread of ideas, Heuneburg offers one clear case in point. Heuneburg was enclosed by a defensive wall made from mud bricks – common, to be sure, but some modern-day historians consider the concept for the bricks highly interesting. The foundation was of local limestone, on top of which, as Cunliffe describes,

> the wall was built of carefully squared mud bricks of regular size and shape. This alien structure, so clearly the concept of a Greek mind, and presumably overseen by a Greek architect, is a striking example of the willingness (and ability) of the local élite to adopt totally foreign modes of expression in the interests of displaying status and power.[37]

Such a flow of concepts and ideas, plus artefacts, such as the thousands of amphorae which have been found all over Europe, add up to vivid evidence of the incredible extent of early trade.

Now, with knowledge of the three interconnected facets of Celtic culture, and armed with the physical artefacts as well as the circumstantial evidence, we can summon everyone to the parlour and finalize our mystery. First up is trading patterns. We know that there were goods, services and ideas flowing from the Mediterranean north around both ends of the Alps and multiple items of trade flowing south. This eclectic mix evidently inspired local innovation in all manner of items and activities. Second of our three facets is the Celtic nobility's love of wine – and their servants' probable dislike of lugging amphorae around – and the common man's love of beer, which needed both small and large containers for fermenting and storing the brew. Combine these with the third aspect – Celtic technology, tools for fine woodworking and the ability to bend wood – and someone thought of a better way to move and store these liquids. The evidence is extremely strong for a nexus of all the elements required to make the long jump from tapered, multi-piece pails and buckets to enclosed barrels: the extensive wood resources; excellent tools with which to work those woods; the exchange of ideas to keep the creative juices flowing; and a love of wine and beer that kept these issues in the forefront of the craftsmen's minds, pushing them to improve their wooden implements. The few barrels which have been found in bogs, well shafts and tombs, indicating that the Celts were the first to have barrels, provide further evidence suggesting that this culture was instrumental in their development.

# FOUR

# Romans: Employing the Barrels for Trade

The townsmen fill barrels with tallow, pitch, and dried wood;
these they set on fire, and roll down on our works.

Julius Caesar, diary entry, *On the Gallic War*

 Roman scribes and artists first documented the use of wooden barrels in the third and second centuries BC. During that period, the Roman Empire was slowly expanding west and north, overwhelming and conquering the indigenous Celtic tribes inhabiting the territories in what are now France, Spain, Portugal, Germany and Britain. Its legionnaires pushed outside the conventional supply-chain routes in use around the Mediterranean, while, as Norman Davies notes, 'Eager Roman merchants had always marched on the heels' of these armies.[1] These forays into unfamiliar environments – from the warmer Mediterranean climate to the colder north – forced modifications in some traditional foods, like wine and olive oil, and to the containers, such as amphorae, in which they were transported. Ironically assistance in adapting to these changes came from the Celtic tribes of northern Europe – the very people whom the Romans conquered. The Celts weren't anxious to share their culture with the Romans, but the latter engaged in what the author Umberto Eco calls 'cultural pillage'.[2] Basically the Romans were superior at arm twisting, eventually squeezing advanced wood- and metalworking skills and technologies out of the Celts.

Oddly, despite several written historical accounts about barrels, and a number of depictions of barrels being used in rather unique ways, there is relatively little known about how barrels were introduced and incorporated within the Romans' daily life. However, even recorded incidents of unorthodox usage such as barrels being employed against Roman troops must have reflected a general acceptance of barrels and their unique properties by this era.

There are a number of stone depictions and reliefs which show barrels in boats. Historians suggest that these contain wine, but they may have transported other goods, since a barrel sitting on a boat's deck in the hot sun would not have improved the wine. Pliny the Elder wrote about 'sanitizing' the vats used for making wine, but he appeared to be referring to ceramic ones,[3] not wooden barrels, although wooden barrels were probably cleaned with many of the same techniques. And several historians mention the barrels used to transport wine from Gaul to Britain via sailing ships, which is more to the point, but few specific details are provided.

Although patchy and lacking detail, these anecdotes indicate that by the end of the first millennium BC wooden barrels were a relatively common container within mainstream Roman trade. The barrel's wide-ranging functions might be considered similar to those of today's metal shipping containers. Witness their now-common, 'extra-curricular' use as construction-project offices, storage units and homes and, in New Zealand, even as an instant shopping mall to replace one destroyed in the Christchurch earthquake of 2011.

In 58 BC the Roman Senate appointed Julius Caesar governor of Cisalpine and Transalpine Gaul and the Illyricum regions: all areas inhabited by numerous Celtic tribes.[4] This was a huge, curving swathe of land that basically ran from the Atlantic seaboard through the Massif Central, north of the Alps, and then down into the regions on the eastern side of the Adriatic. The appointment gave Caesar the means to take control of the many Celtic tribes that were still in fierce opposition to Roman rule. Over the following sixteen years he battled those indigenous tribes with an array of creative military strategies, ultimately subduing them.

One early tale of barrels being used against Roman troops comes from Caesar's diaries and other accounts and concerns the siege of a hilltop village in Gaul with the barely pronounceable Roman name of Uxellodunum. This village, known now as Vayrac, France, is located on the Dordogne River roughly midway between Brive-la-Gaillarde and Cahors. In 51 BC it was one of the last to resist Caesar's legionnaires. Between 53 BC and 51 BC, under Caesar's command, Roman legions engaged in many battles with the numerous Celtic-Gallic tribes of central France, including the Arverni (Auvergne region),

Senones (residing in the departments of Loiret, Seine-et-Marne and Yonne), Bituriges (tribes which stretched from Bourges to Bordeaux) and occasionally the Aeduans (in the present-day departments of Côte-d'Or, Saône-et-Loire and Nièvre), who twice switched sides.

As the battles wound down during the summer of 51 BC, according to historian Stephan Dando-Collins, Caesar's 10th Legion 'and its fellow legions spent the rest of the year mopping up the final resistance'.[5] When the Gallic defenders of Uxellodunum were surrounded by the Roman troops, as a last-ditch effort of resistance they packed small wooden barrels with flammable material – 'pitch, tallow, and firewood' – and hurled them, burning, down the escarpment.[6] As the flaming barrels tumbled down, they ignited the hillside brush. Thinking that the besieged Gauls might try to escape by hiding behind the smoke, Caesar's soldiers attempted to extinguish the fires, only to become targets of rocks, spears and arrows descending from above. Eventually, through diversionary tactics, Caesar's legions were able to subdue the village, cutting the hands off many

Map of the provinces of Transalpine Gaul under Augustus.

of the survivors as a lesson. The legionnaires went on finally to conquer all of Gaul.[7]

The Roman historians writing about Uxellodunum referred to these barrels as *cupae*, and it is from that word that the French, expanding upon the size of the container, derived the word *cuve*, or tank.[8] The Roman cooper, the person who worked with the staves, was *vietor doliarius*,[9] with the French word for a stave maker becoming *doleur*. However, the role of cooper may have been a secondary one for a wheelwright or blacksmith, as there is no listing for 'cooper' under the Skilled Labour section in the Roman emperor Diocletian's *Edict of Maximum Prices* issued in AD 301.[10]

Barrels are described in another encounter with Roman troops – to the legionnaire's benefit on this occasion. In AD 238, Emperor Maximinus, also known as the Thracian since he was born in Thrace, modern Bulgaria, was travelling to Rome with his troops to save his emperorship. Faced with crossing a swollen river at Aquilea, a town in northeast Italy near Udine, he found the bridge destroyed. The river was too high and swift to ford, even on horseback. To solve this problem, his engineers creatively built a pontoon bridge, using discarded wine barrels left by the fleeing Italians. The new bridge was a success. However, once across the river, Maximinus found the town of Aquilea locked tight and, although he lay seige to it, he couldn't break the defenders' hold. When his army ran out of food, he foolishly berated his officers and men for their predicament. On empty stomachs, that was the last straw. His men could no longer accept their general's rather harsh, authoritarian rule. His opponents, buoyed by the now-common disgust with the general, killed him and his son, taking their heads to Rome.[11]

These two examples of the early use of barrels are found in almost every history about cooperage. The most important thing to take away from these episodes is that the people involved – the defenders and the military engineers – were able to incorporate barrels into functions beyond their standard use as containers for provisions and liquids. This, I believe, indicates that the barrel must have been considered a relatively common container by the Romans. Unfortunately this is an aspect of daily Roman life about which there is little information.

We can reasonably assume that barrels were used to store and transport many commodities, although, in the early days of the Roman Empire, what they contained other than wine remains rather elusive. There is scant evidence even to indicate exactly when Roman and Gallic vintners started using barrels. Wine growers living in Italy and around the Mediterranean, who had used amphorae for many years, may not have been comfortable changing over to barrels. Besides the common tendency to be against change, perhaps the barrels were not as practical for several reasons: amphorae were already readily available; with the drier summers, and relative lack of climate control in cellars and warehouses, causing the staves and headers to dry out, it may have been difficult to keep the wooden barrels tight, further oxidizing the wine or causing leakage; and the use of Alpine silver fir to make barrels might have meant that pitch needed to be added on the interior to seal them,[12] thus adding an undesirable flavour to the wine. Finally, since barrels were still rather uncommon in the trading system, they might have been more expensive, on a per-litre basis, than amphorae.

However, when the Romans introduced wine grapes to France, southern Germany, Austria and Hungary, they were doing so in a region which had a shorter history of grape growing and thus weaker and less ingrained winemaking traditions. Since these northerly peoples were probably already familiar with barrels being used for other commodities, they were no doubt more accepting of barrels as they became available within the general trading scheme of these regions. Barrels, like amphorae, could keep oxygen out of the wine, but had the practical advantage of being easier to move (rolling the barrels instead of lifting the amphorae, as previously mentioned) – and the barrels had a far greater capacity for their weight.

The records are also not clear as to whether the Romans and their contemporaries were using barrels just for the transportation and storage of wine or also for ageing it. Certainly there is some ageing occurring during the transportation time frame as the wine travels from the winery to the consumer. But were the vintners allowing wine to remain in their cellars for one or more years? And was ageing being done in the négociant (trading house) cellars, or even in the cellars of the Roman consumers?

BACCHVS.

8

Bacchus, the Roman god of wine, is shown astride a barrel in
a 16th-century drawing. Note the head brace, and the small
racking bung at the bottom centre of the barrel head.

An educated guess says yes to all, once two dynamics had come together and evolved: growing the grapes further north and making barrels from flavour-enhancing woods such as oak, or at least flavour-neutral woods such as chestnut. However, while and when barrels of silver fir were in use, ageing wine in barrels was most likely accidental.

If the vintners around the Mediterranean did not age wine in barrels, they aged at least some wine in ceramic containers. This was obliquely noted by Pliny when he suggested emptying the containers at the 'rising dog-star' or start of the hot season; the wine would then have been in a container for about nine months. Another bit of evidence comes from the *Edict of Maximum Prices*, which notes separate prices for both 'young' and 'aged wine' – ordinary wine was 6 price units per *sextarius*, a standard Roman quantity of 564 millilitres, while aged wine was 20.[13]

As the Romans introduced grape growing further north, and with the probable use of barrels in those locations, some ageing probably occurred on the occasions where wines were delayed in shipping after fermentation finished. When exceptional vintages made full-bodied wines, ageing may have been practised. This could have occurred when the vintner felt that a bit of extra time in his cellar might improve the wine, especially so if he could realize a higher price.

Researchers note that in the Bordeaux region the common practice of prolonged skin contact during fermentation seemed to have begun in the Middle Ages, as did the use of the wooden grape press.[14] However, the press was in regular use centuries earlier in the more established wine-growing regions. Certainly by the Middle Ages wooden barrels were in common use, as was ageing – if not in the vintner's cellar, at least in the trader's. Depending on where and when usage of the big wooden basket presses first started, perhaps barrels came into use around the same time for ageing the resulting tannic wines.

While some Mediterranean shipwrecks contained up to 5,000 amphorae, ships plying the Atlantic probably carried barrels numbering in the tens to hundreds. As it awaited shipment, wine would sit in the trading warehouses. And again, upon arrival at the destination, in all probability it was not dispersed immediately, so quite likely remained in storage for some period of time. Ageing, either accidentally or on purpose, would have occurred.

Finally, some of the Roman estates had wine cellars.[15] Whole barrels of wine were purchased by at least a few wealthy Romans. It is possible that instead of transferring the wine to ceramic pots, they allowed it to remain in the barrel, stored in their cellars. They may even have understood that wine evaporates from the barrel and must be topped up occasionally.

Roman artists working in stone did record barrels being used, although in most cases it was not clear for which product. The Roman emperor Trajan, after subduing indigenous tribes in what was Dacia (an area to the north of the Danube River, roughly equivalent to modern-day Romania), erected a column in Rome to commemorate his armies' exploits. Completed in AD 113, hundreds of figures portraying his troops' deeds are sculpted in a relief which marches up the column in a helix pattern. At the bottom, within several of the panels, are scenes of what appear to be preparations for the war. Several barrels are lined up on the shore. In one small river boat, two men are seen manoeuvring a barrel into place. Another small boat has three barrels stacked up on the deck. A third boat appears to have a load of bags, perhaps filled with grain or flour.[16] That the barrels contain wine is a reasonable conjecture, but not confirmed.

Another Roman-era 'barrels on a boat' image is depicted in a stone bas-relief found at Cabrières-d'Aigues, a village of ancient Gaul set in the hills of the Durance River's northern floodplain. The Durance was utilized as a conduit for inland trade, along a corridor in southeast France and western Italy, just south of the Alps. The relief shows men pulling a boat, via ropes, from the towpath at the river's edge. Additionally, near Trier, on the Mosel River in west-central Germany, a Roman-era stone boat was excavated, and it too had barrels prominently displayed on its deck.[17]

With the Romans shipping wine to all parts of their empire, it is not surprising to find barrels in Britain (as already mentioned, the amphora had been the container of choice up until the second century AD, after which its usage declined rather abruptly as it was replaced by the barrel[18]). The barrels that were used as well linings in Roman Britain demonstrate how barrels can be seen in a much

A bas-relief section from Trajan's Column, AD 113, Rome. The barrels on the boats have provisions for Emperor Trajan's war with the Dacians.

wider way than simply as containers, and this was also one of the few ways in which a barrel could be accidentally preserved. Barrels in wells have been found in major excavations at Colchester and Aylesbury, north of London, and in Silchester, to the south.[19] Historian Joan Liversidge surmises that the barrels originally contained wine shipped from Bordeaux.[20] Given that they were made of silver fir, if the vineyard source was in southern Gaul, perhaps the wood came from the fir forests of the Pyrenees. Other historians note that some of the wine shipped to Britain came from regions along the Rhine, perhaps shipped down the river and across the Channel. And wine from France and Italy making its way to Britain was also shipped, where possible on rivers, or travelled overland after the Romans improved the road system, before crossing the Channel.[21] The silver fir for barrels from northern France and eastern Germany would probably have been sourced near the Alps. The wood of the silver fir, commonly termed Alpine silver fir or European silver fir, is stronger than pine, and is somewhat like the Douglas fir of the western American forests. It can be bent slightly, but not as much as oak, and would have imparted a slightly resinous flavour to any product placed within it.

A final unique Roman use of the barrel was to collect urine. In Pompeii and other Roman cities, barrels and clay pots were placed on street corners, strategically near bars and inns, to collect urine for fullers, the craftspeople who cleaned wool and cloth. With its ammonia content, the urine was used to remove the fat from wool. Other urine users were the dye-makers, launderers, metalworkers and leather tanners, all of whom had to pay a tax to obtain the urine. This tax was imposed by Emperor Vespasian during the late first century AD, and from his name comes the Italian word for urinals: *vespasiani*.[22]

All these depictions and stories of barrel usage seem to indicate that they were slowly becoming mainstream containers. It is not clear whether they were utilized first for storing and transporting wine or for other commodities, primarily since historians and archaeologists do not fully understand Roman and Celtic barrel usage. But given the sophistication of Roman society and culture in so many areas, the evolutionary dynamics and processes of incorporating the

This Roman-era stone boat of *c.* AD 220 was unearthed in the village of Neumagen, located in the Piesport wine region of the Mosel River. Wooden barrels containing wine are prominent on the deck.

wooden barrel would not have been significantly different to what we see today with the introduction of new and different types of product packaging.

## Barrel Development in Other Cultures and Civilizations

The wooden barrel – that is, a cylindrical container with bowed sides – was developed only in Europe. Other regions, including Asia, independently developed wooden buckets and tanks with multi-piece but straight sides.[23] It wasn't until the eastward treks by the Crusaders, and then the exploration of the world via early sailing vessels, that European-developed barrels were introduced into Asia and the New World.

Like the civilizations around the Mediterranean, China had been utilizing ceramic jars and pots and buckets of various designs and

materials. By AD 1000, drawings show that enclosed wooden tanks – termed *taru* – were being built to hold and transport sake.[24] Unlike the barrel, they were straight-sided, but did taper. With rope handles for carrying, their size would have been limited unless the handles could be attached to lifting poles supported by several men. Or perhaps they were carried only short distances, say from a warehouse to a wagon, cart or river barge.

There is no evidence that the early peoples of North, Central and South America developed barrels. Certainly some used hollowed-out logs for buckets, drums and dugouts and canoes, and birch bark was in use as a container and for boats. But despite the abundance of forests and excellent woods for cooperage, there doesn't appear to have been any use of wooden barrels.[25] It appears that, while the need for barrels may have existed, neither the nexus of metal tools required for precision woodworking, nor the development of wood-bending skills, had occurred. R. F. Tylecote notes that South American cultures advanced only to the bronze stage; there was no iron smelting until the Europeans arrived.[26] Additionally, with no sizeable beasts of burden – horses or oxen – in North or South America until the arrival of Europeans, moving barrels any great distance over land would have been difficult.

The barrel development story, or lack thereof, in Africa was similar to that in the New World. Interestingly, despite the early advances in multi-piece buckets noted in Egyptian tombs, further development into a barrel did not seem to take place.

# FIVE

# Middle Ages:
# A Surge in Barrel Use

By the time of the Crusades [starting in 1095] wooden casks were the
standard means of transporting all manner of liquids and provisions,
and the cooper was coming into his own as one of the foremost
tradesmen, particularly in coastal and riverside towns.

Kenneth Kilby, *The Cooper and His Trade*

Wine imports could be naturally affected every year by bad harvests,
plague, war, and even by the effect of the weather or piracy upon
voyages. Apart from prolonged war, these factors would have had
a short-term effect on the pattern of trade.

Catherine Pitt, 'The Wine Trade in Bristol in the Fifteenth and Sixteenth Centuries'

 The year 1985 found me and my wife travelling through
Europe visiting wineries and cooperages. In Bordeaux
the particular cooperage we were seeking was located
deep within the Chartrons Quarter, a warren of old stone wine
warehouses adjacent to the Garonne River. On stepping inside, it
took a few moments for our eyes to adjust to the dim, hazy atmos-
phere. The walls, blackened by years of smoke from the bending
fires, absorbed what little illumination entered from the few dirty
skylights. Sawdust and shavings littered the floor, and the air was
filled with the rich aroma of toasted oak. We were met by the owner
and, over the constant clang of hammers pounding the hoops,
were quickly passed on to his wife, who ran the business. It was the
classic small-business case of an owner who just wanted to engage
in a craft, in this case making barrels, and a wife who wanted to know
where the money was coming from. Letting him make his barrels,
she took over the accounting and looked after the customers.

As we observed the barrel-making activity, we could see men
at each machine, often with another awaiting his turn. Madame
told us that her coopers worked on a piece rate, a common method
for maintaining productivity. This meant that each man built and

completed as many barrels per day as possible. However, with only one machine of each type, additional pressure to produce quickly was created by knowing that someone else was impatiently awaiting use of the machine.

This cooperage business no longer exists in that building; typical of many French companies, it has since moved outside the city into a modern facility, with the son and his children now managing the business. On reflection, with the exception of the machines, the image of that Bordeaux cooperage we observed in the 1980s was probably not that different from those seen in drawings of fourteenth- to sixteenth-century European cooperages. And, quite possibly, things really hadn't changed too much even from Roman-era cooperages.

## Bordeaux

Bordeaux is a good place to begin our examination of the use of barrels as the Roman era ended and the Europe we now know began to emerge. Even before Julius Caesar fought his way through Gaul, Burdigala, as Bordeaux was known to the Romans, was a trading centre for wine.[1] However, at that time the wine was not grown in Bordeaux but sourced from Italian vineyards, primarily around Etruria and Campania, in west-central and southwest Italy respectively.[2] It was packaged in amphorae, secured for the long journey with a wooden plug sealed with wax. Some loads were sent overland via Marseilles, while others were shipped across the Mediterranean directly to Narbonne. From Narbonne, a consignment would be conveyed by land or rowed in boats up the Tarn River, to Toulouse, and then barged down the Garonne River to Bordeaux. Merchants in Bordeaux sold the wine on to the Atlantic markets – western France and Britain. Even in this early era, Davies notes that 'Roman Britannia was fully integrated into the imperial trading systems'.[3] Italy to Britain was a long trip for a fragile commodity, but, surprisingly, not atypical of the long-distance trading taking place at the end of the last millennium BC and the beginning of the first one AD.

With the Roman takeover of Gaul, vines and winemaking skills were introduced into the southern regions. Eventually, plantings slowly made their way north and west, as farmers and landowners wanted to propagate the grapevines and make money from the resulting wine. Once fighting the Romans ceased, the Bordeaux merchants encouraged plantings and winemaking in their own region. Tired of paying duties to Toulouse and Narbonne for the wine coming through those cities, they believed they could make more money by producing their own wine.[4] Slowly, the previous dependence on wine from Italy, and also from Spain, decreased, as the newly planted vines in the Bordeaux region came into production. And by the thirteenth century, 'almost all Gascons, from the Archbishop of Bordeaux downwards, were involved in the wine trade', as Anne Crawford notes in her book about the Vintners' Company, a London-based guild and major trading partner with Bordeaux.[5]

The early Bordeaux vineyards were thought to be planted with grape varieties such as Allobrogica, an early scion of Syrah, and Biturica, possibly a forerunner of Cabernet Sauvignon.[6] The Allobrogica came from east of the Rhône, so it was already fairly well acclimatized to the cold Bordeaux winters. The Biturica may have been brought from Italy or Spain.[7] Without the apparent cold-hardiness, Biturica needed careful siting in warmer microclimates.

Around the first century AD the merchants of Bordeaux, desiring further control of their wine-trading network, set up a factory to produce amphorae.[8] However, this enterprise was relatively short-lived as the continued use of amphorae, particularly for overseas shipments to Britain, did not last long. For reasons that are not entirely clear, in about AD 250 the use of amphorae ceased rather abruptly as barrels took over and became the primary wine-shipping containers.[9]

In the case of trade between France and Britain, the shift from amphorae to barrels may have been partially influenced by the type of ships that plied the Atlantic between mainland Europe and British shores.[10] In 1982 what is believed to have been a typical merchant ship of the period was discovered buried in the mud of St Peter Port harbour, on the Isle of Guernsey. The *Asterix*, as she was called, dated from about the second or third century and was a sailing ship of some 25 metres in length by 6 metres wide, with a freeboard of about 1.5 metres.[11] Built with thick oak planks, fully decked and with a curved bow and stern, she was capable of weathering the rough Atlantic seas in order to transport cargo between ports in Britain, Ireland, the Netherlands, France, Spain, Portugal and possibly as far as the Mediterranean.

Ships like this were substantial, capable of carrying at first many amphorae, and later many barrels.[12] The *Asterix* is estimated to have had a cargo hold of about 200 cubic metres.[13] Conservatively estimating 2.22 cubic metres per barrel (using the size of the theoretical standard barrel of the era, the 900-litre tun), the ship could conceivably have carried as many as 90 barrels of the tun size, although typically the cargo might have been a mix of commodities and containers. The weight for that amount of barrels, full of wine at 997 kg each, would have been about 90 metric tons. That weight would equal approximately 2,500 amphorae, which was also about an average ship's cargo.[14] A full load of wine in those 90 barrels, at today's value of, say, $2 per litre wholesale, would have been the equivalent of

A scale model of the *Asterix*, a ship which was in use prior to AD 280 for transporting wine and other commodities to and from Europe and the British Isles.

$162,000. During that era wine was shipped in bulk, with little or no mark-up for sales or marketing – the packaging *was* the barrel. The majority of ordinary people at the time drank beer; wine was generally intended for the upper class and the clergy, and would have been priced accordingly. And despite the *Asterix* being, as Bob Dean notes, 'a most sophisticated vessel, the end-product of an evolutionary design process surely dating back many centuries',[15] hauling barrel-loads of wine, or any product, across the open ocean was a risky business, but rewarding when successful.

As opposed to the flatter-keeled ships which plied the Mediterranean, those Atlantic-voyaging vessels had more of a V-shaped hull, designed to cut through the heavier seas. However, their keel was not so deep that it deterred them from navigating the shallow estuary waterways of end-of-the-tide ports such as Bordeaux and Bayonne or London and Bristol. The cargo hold would also have had a V shape. Within these holds barrels nested better than amphorae, as already mentioned, and may have been a significant factor in the shift from amphorae to barrels.[16]

In the ensuing centuries, Bordeaux continued to be a major exporter of wine,[17] with barrels consolidating their position as the primary storage and shipping container. Strangely, first-millennium cooperages are not mentioned in the historic literature. It is possible that the earliest cooperages were not separate workspaces, but adjuncts for the wooden articles being made by carpenters, blacksmiths or chariot and wagon makers, as the skills were similar. However, as the demand for their work increased, the craft of cooperage developed into its own specialized trade. Evidence for this comes from second- and third-century funerary steles in the Bordeaux area depicting, among other specific trades, *tonneliers* or coopers.[18]

But these coopers, and the shops in which they worked (which we will call cooperages, although they were not full-blown cooperages until several centuries later), did not simply appear out of nothing. A whole industry was necessary to produce the cooperage products: sawyers to fell the trees; the *merrandier*, who splits out the rough staves from the logs; the *doleur*, who trims and planes the staves; the 'cousin to the stave maker, the hoop maker', as Jean-Marc Soyez termed the men who produced the willow and chestnut hoops (in the many centuries before the common use of steel hoops);[19] and, finally, the cooper to assemble all the parts to produce the barrel. Sometimes these activities were carried out by the same individual; however, as demand increased, the jobs became more and more specialized.

Today's cooper primarily just makes barrels, either for wine or whiskey. But the independent coopers and their associated craftsmen of 2,000 years ago, and even up until just a few hundred years ago, made an assortment of wooden pails, buckets, pitchers, churns, funnels, tubs, barrels, tanks, vats and other wooden objects. The buckets and pails were required to carry water from wells, springs and rivers. Pails, churns and buckets would have been needed for hand-milking cows, sheep and goats, and the more elaborate and specialized ones were used to process the milk into butter and cheese. Various styles of wooden pitchers were used at the table to pour water, wine, beer or cider.[20] These pitchers have sides which were bent to produce the curved shapes we are accustomed to seeing in ceramic and glass water jugs, the construction of which would have been similar to bending wood for the larger barrels. Small tubs would have been used to mix flour for bread and larger ones were necessary for crushing the grapes. There are numerous historical prints and tapestries depicting people standing inside the tubs stamping on the grapes with their bare feet. Other tub-like containers were required for pickling fruits and vegetables, making beer and cider, salting meat, processing the wools and fabrics for clothing, washing clothing and tanning leather. And as more and more wine was made, larger tanks were needed to store some of it.

Whenever a liquid needed to be contained, and the choice of a ceramic or wooden container was required, the vast, northern European forests of mixed species provided ample opportunity for wooden buckets, tubs and barrels to meet such requirements satisfactorily. And for many of these wooden containers, oak was a primary material. Jean-Paul Lacroix estimates that the proportion of oak used in cooperage was around 45 per cent in the second century AD, rising to 100 per cent by the third century.[21] Lacroix further notes that the 'strength and durability of oak, as compared to softwoods, was a critical asset'.[22] These significant

French peasants in this 19th-century etching are playing an early version of Palets. Here, the hole in the top of the barrel serves as a jack, as in Boules or Pétanque, with each player attempting to toss his small disc nearest the hole.

factors favoured wooden, and more importantly oak, containers over amphorae for everyday use and for containers used to transport various commodities.

Despite a growing demand for wooden containers, coopers were subject to the same economic ups and downs as the Bordeaux wine merchants and other tradespeople of the period. With the Roman Empire slowly disintegrating, starting in the first century AD and extending to its demise (of its western lands at any rate) by the fifth century, there was a continuing disruption to the wine-supply chains. Roman wine traders and négociants were displaced by regional princes, bishops and monasteries, in part as a power grab with the break-up of the imperial infrastructure, and also to ensure wine for their religious services.[23] As these new political entities came to power, they then attempted to tax wine as it was moved about on the rivers and roads, which to various degrees impacted on vintners and, by extension, coopers.

Other disruptions occurred when Germanic tribes pirated coastal shipping during the second to fourth centuries, periodically impacting Bordeaux's lucrative trade with Britain. And although the trade to Britain was improved by the marriage of England's Henry II to Eleanor of Aquitaine in 1158 (perhaps aided by the marriage gift from Chancellor Thomas Becket: two chariots loaded with iron-hooped beer barrels),[24] later it would be the Hundred Years War (1332–1453) between France and England, as well as plague and further external and internal wars, which dealt serious and periodic blows to the economic and cultural landscape.[25]

Further trade struggles for Bordeaux's wine industry arose from the development and growth of vineyards throughout northern France, around such areas as Paris and beyond. They were serious competition during the early stages of their growth, but over time became less so due to their unproductive sites – eventually only certain northern regions were found suitable for grape growing: those prevailing today which we see primarily in the northern Burgundian, Alsatian and Rhine regions. Another problem encountered by Bordeaux merchants was that when they were prevented from selling into Britain, either through certain laws or crippling taxes, vintners from Portugal and Spain slipped in to fill the void.[26]

Not all the wine shipped from France arrived, or arrived in good shape. Some ships sank; others were pirated. Records indicate that of the wine that did arrive in Britain, often as much as 10–20 per cent was 'corrupt' – either having gone off during shipping or because it hadn't been particularly good to begin with. An equal percentage was lost to leakage or barrel ullage (evaporation).[27]

Unlike today's overnight cross-Channel ferries, or the two-hour train trip under La Manche, a sailing voyage bringing wine to Britain could take weeks or months, what with the time involved awaiting appropriate winds and tides and stops at various ports for supplies and additional cargo.[28] For example, exiting the Garonne river by sail required an outgoing tide along with an easterly breeze. Ships sailing to Britain could encounter difficult seas while also having to avoid pirates and navigate around the westward thrust of the Brittany peninsula. And the journey might be delayed by another wait for an incoming tide and easterly wind prior to entering the Thames estuary. There could be further delays awaiting London's bureaucratic formalities, which occasionally meant that another three days would pass before the wine could finally be offloaded at a London pier.[29]

The strength of Bordeaux's early wine exports was not insignificant. By the fourteenth century it was exporting some 700,000 hectolitres annually,[30] the equivalent of almost 8 million cases, in some 77,000 barrels. It was in the fourteenth century that typical loads of 20,000 tuns (the large 900-litre barrels), or the equivalent thereof, were shipped annually from Bordeaux, and from its southern neighbour Bayonne, to Bristol alone – a distance of over 900 km by sea.[31] Unfortunately export figures for Bordeaux at the beginning of the first millennium are not available, but a comparison can be drawn from estimates of first-century shipments between Italy and Gaul, which are in the order of 120,000 hectolitres (or 1.3 million cases) of wine annually. Today Bordeaux's total production is over 2 million hectolitres – some 22 million cases.[32]

Stemming from the Romans' extraordinary zeal for standardization and taxation, one term, still very much in today's lexicon, resulted directly from the use of cooperage during these years of oceangoing wine trading. The current word for a ship's capacity – 'tonnage' – originated in that early part of the first millennium. The term evolved from the French word tun, the name for the large barrels of 900-l capacity, and from how many of these could be loaded (and thereby taxed) onto a ship. Although the term has been shifted to refer to the capacity of the cargo space, it is still the standard by which we measure and compare ships.[33]

From my own barrel-handling experience, and despite numerous references in the literature (which may just be repeating each other), while these 900-l barrels may have been the standard measurement for shipping wine, I find it hard to believe that they were actually the most common barrel size in use. These barrels would be almost 2 metres tall by 1 metre in diameter and full of wine would weigh 920 kg – almost a ton! I have rolled puncheons of 500–600 litres, which are difficult enough even when empty. Despite my previous praise for the ease with which barrels could be rolled up ramps, even with two or three men, plus others pulling on a rope around the 900-litre barrel, rolling a 920-kg barrel would be arduous and push the boundaries of what normal labour can achieve. When these large tuns were loaded into the ships by a block and tackle attached to the yardarm of a mast, as they most likely were, positioning them within the hold of a moving or rocking ship would be no easy feat. I suggest that the majority of the barrels shipped from France to Britain were not 900-litre tuns, even though that size was the stated measurement for the ships. I believe that the majority of the barrels were the 225-litre size – an even four of those casks making up the full 900 litres.

Several notes in the literature support this theory. Aubin, Lavaud and Roudié, in their book about the history of wine in Bordeaux, wrote that the early coopers were kept busy making new barrels for each year's harvest as the previous ones were used to ship the wine to Britain.[34] However, considering the effort required for vintners to move 900-litre barrels out of a cellar, on to a cart and down the primitive roads to the Bordeaux port suggests that such a task would have been extremely difficult. Aubin and his colleagues concur, proposing: 'It is considered, therefore, that the average capacity of the barrel is 200 to 225 litres.'[35] While the ports would have had numerous hands for the labour required to shift and transport barrels, the vintner, trying to get his wine to the port, most likely did not.

Second, the Asterix (see p. 62) was 'fully decked'. This would have been essential to avoid swamping in heavy seas. So any cargo of barrels would have been lowered into the hold via a midship hatch and then wrestled into position. I find it difficult to believe that the crew could, or would even attempt to, roll one of these large barrels over others already placed within the hold to double-stack the larger barrels, which would have been necessary to carry the indicated loads.

As a reasonable alternative, if 900-litre barrels were routinely shipped with each vessel, perhaps they were just placed in the hold, near the hatch, with the balance of the hold space filled with smaller barrels, which would have been easier to manoeuvre. This is further supported by Pitt's research into the Bristol shipping records, which provide historical documentation recording the number of tuns received on various ships.[36] As the record of a typical ship, the Facon arrived in Bristol from Bordeaux on 2 March 1574 with

'50 tuns' of Bordeaux wine. The record further breaks down these 50 tuns to 35.5 that held good wine, 6.25 tuns containing corrupt wine and 5.75 of tuns that had leaked. As these tuns are delineated with quarter decimals, the figures very likely represent actual 225-litre barrels. Further, these figures add up to only 47.5 tuns. Pitt noted that the remaining 2.5 tuns were 'other' cargo – again adding to the idea that these could have been smaller barrels or other items of freight.

Selling wine has always been an erratic business, as today's winemakers well know. In the Middle Ages another factor contributing to this cyclical nature, with its consequent impact on the need to produce barrels, was the fickleness of the wine-drinking public. In those early days, when beer was the main beverage for the majority of people, wine was still new to many northern Europeans, especially the emerging bourgeois class. The Celtic nobility had acquired a taste for it, but it was most likely out of commoners' price range. Like many inexperienced wine drinkers today, some probably did not care for the rich, full-bodied wines being grown in Bordeaux, although this heavier style, with its higher alcohol content, was probably favoured by many of the vintners because they were more confident that it could be transported to their customers without serious deterioration.

In addition to the British market, the emerging commercial strength of the Dutch provided a powerful but short-lived marketplace for Bordeaux wine. In general the Dutch preferred sweet white wines, not the typical, heavier Bordeaux styles. Some of Bordeaux's neighbours to the north, around Cognac, who were growing white grapes, capitalized upon this demand.[37] With the Dutch taking a financial interest in some French grape estates and wine-processing operations, they added another twist – distilling the grapes. By 1725, this activity had established the famed Cognac and Armagnac brands. Distilling grapes added worth to otherwise less valuable grapes and surplus crops, reduced shipping costs by lowering the quantity of liquid needed (more bang for the buck, if you will) and ensured that the wines travelled better due to the higher alcohol levels.[38]

The trade with the Netherlands impacted on Bordeaux vintners, merchants and coopers in several ways. The Dutch were efficient

A typical early wharf scene with whale oil barrels and casks, New Bedford, Massachusetts. The E stencilled on the barrels indicates that they are empty. In the background, the whaling ship *Greyhound* has its whale boat perched athwart its stern.

merchants and did not send out empty ships. While wines and Cognac travelled north, filling the ships with other money-making products for the return trip south to pick up more of the French cargo was also a requirement. One of the products that filled the holds was rough barrel staves. The wood was sourced from northern Germany, Poland and from the eastern Baltic seaport of Memel, now Klaipèda, Lithuania. Memel oak continued to figure prominently, even up to the nineteenth and twentieth centuries, as a source of staves, which were sent to Bordeaux coopers and cooperages. But those cooperages who sold to vintners in the Cognac region saw their orders decrease, since fewer barrels were required when the wine was distilled into higher-proof alcohol. As a result of these various difficulties for the Bordelais cooperages, they were able to pressure the Bordeaux city fathers into requiring upstream vintners to ship their wines in barrels smaller than the traditional 225-litre ones used by Bordeaux vintners. Customs agents taxed the wine on the *number* of barrels, not the volume, that passed through the city.[39]

## London

The major recipient at the end of the Bordeaux wine hose was Britain. Early on, the Roman and Bordeaux region merchants were the prime movers, but by the fourteenth century wine merchants in London and Bristol had taken control of the business and coordinated the shipments.[40] Unheralded, but providing the essential container services for the trade, were the coopers on both sides of the Channel: French coopers building barrels for exporting the wine and British coopers who repaired them and kept them sound after the cross-Channel journey, as well as making barrels for shipping other commodities back to France.

By the fourteenth century the city of London was importing a third of Britain's wine – several million litres of wine per year, which in itself was a significant portion of Britain's entire import trade.[41] Generally the wine came in two major shipment waves, each involving hundreds of ships. Vessels carrying the just-fermented wines sailed in the autumn, akin to the Beaujolais Nouveau affair, while the wines that were allowed to mature, and had been clarified by racking, were shipped early the following year.[42] The Bordeaux coopers would be busy, especially during the summer months, preparing barrels for the harvest. At the same time, their London and British counterparts made barrels for salted fish, cheese, butter and meat to be shipped to the Bordeaux wine merchants, who in turn would sell those items, along with English wool and other British exports, into the Gascon market.[43] The used wine barrels should not have been reused for the British export commodities as they would adversely have impacted the products, especially the cheese and butter. However, some staves or heads were certainly reused occasionally, and perhaps a few wine barrels were slipped in unbeknownst to the merchants.

Imagine the London dock scene with the arrival of a fleet of wine-delivering ships. With several hundred ships like the Asterix crowding the Thames, such a scene, which was played out for hundreds of years, was designed to be orderly but was undoubtedly chaotic. The designated areas for unloading wine were the old Garlickhithe and Queenhithe wharfs, a few blocks downstream from London Bridge and the site of Billingsgate fish market until 1982.[44] Located near this receiving area was Vintners' Hall, the guild centre of the prestigious Vintners' Company, the merchants who supervised the massive importation.

In the reign of Henry II (1154–89), ships had to wait before unloading while sheriffs and the king's chamberlain tasted their way through the barrels aboard each ship – choosing one from before the mast and another aft – to decide which would be selected for the royal cellar.[45] In those days the king had the right of prisage, the tradition by which he could take a percentage of a cargo and purchase an additional quantity at a set price for his household and military needs. Prisage predates the more modern system of paying customs and duty taxes.

Once the king's casks had been designated, the ship was allowed to pull alongside the dock or moor nearby to offload their cargo. A ship's crew would hoist the barrels and tuns out of the holds to the awaiting wine porters, the paid servants of the wine merchants. If the ship was wharfside, the porters would roll them down ramps to the dock. From ships moored in the river, the porters would ferry the barrels to the wharf in lighters: small, short-haul barges. The barrels would first be placed in the vintners' wholesale warehouses. Then, upon sale, the porters, for set fees, would distribute the barrels to the royal cellars, to taverns and to individuals who could afford to purchase whole barrels. For destinations only a short distance away, barrels were rolled up the cobbled streets, and for delivery further afield they were loaded onto horse-drawn drays and wagons.

With senior porters, and other assistants to the wine merchants, directing the flow of hundreds of barrels, the labourers grunted and strained to roll the heavy barrels directly onto awaiting carts or place them in rows for future distribution. The shouts of the men, the rumbling of the barrels on the stones and the stomping of the impatient dray horses all would have added to a cacophony of sounds and activity. Inclement weather, especially in a drizzly London winter, must have made the whole event even more muddled.

Invariably some barrels leaked during shipment and would need repair, so amid all this dock activity were several of the London coopers, tools in hand and ready to attend to any problems. According to the hand-manufacturing standards of the day, leakage was acceptable, or at least tolerated; it would be totally unacceptable by today's standards. London coopers would plug wood leaks with small spiles, conical or wedge-shaped pieces of a softer wood which could be tapped into a hole made by an awl or chisel and which would expand upon contact with the liquid, closing the pores in the wood grain. If the leak was between two staves or two headboards, pieces of waxed string or reed could be pounded into the gap. The coopers' repair services were also required while wines rested in London cellars awaiting sale and shipment to taverns, upper-class houses or church facilities.

Information about the earliest London coopers is gathered from the limited excavations, a few artefacts and a smattering of historical records. Old documentation describes Cooper's Row, a small street besides the city wall that which was home to many craftsmen. Strategically located for the coopers' business, it was just a few blocks from the wine vaults near the Tower of London and the wine and fish trade at the Thames docks.[46] Twede suggests that coopers located their shops at this and other industrial sites throughout the city, congregating near the docks, breweries and flour mills, in order to provide packaging services – supplying new barrels and repairing older ones – for the products of those industries.[47] In medieval Europe it was not unusual for a street to take on the name of a trade, in this case 'cooper' or occasionally the variation 'cowper', as a reflection of the area's original inhabitants.[48] A small remnant of ancient London's Roman-era fortress wall provides the only historic link to Cooper's Row; the street itself is now home to several trendy shops, hotels and bars, with the cooperage link remaining in name only.

London's original, fifteenth-century coopers' guildhall was on Basinghall Street, near, but not to be confused with, the Guildhall, London's first city hall. Unfortunately, over the years, confiscations, fires and Second World War bombing took their toll, and in 1958 a new guildhall for the London coopers' guild was purchased at 13 Devonshire Square, a late seventeenth-century merchant's house built on the gardens of the Earl of Devonshire's townhouse, just a few blocks from Cooper's Row.[49]

On Cooper's Row each cooper would have had a small shop, facing the street, which also served as a showroom.[50] The cooper's living quarters were usually above his shop. If he had an apprentice, the young man would have slept in the shop and eaten with the family. Most of the cooperage bucket and barrel products were built to order, but some common items might have been made for selling from a booth during one of the city's market days.

Research suggests that medieval coopers in and around places such as London and Bristol were closely associated with the wine-selling operations, which were major barrel users, and that such coopers were probably relatively independent craftsmen. The first situation has remained so to this day, while the second has changed slowly with the advent of guilds and, later, the larger, multi-man cooperages.

In those days, quality control was in the hands of the craftsman; it depended upon his pride, his skill and the sharpness of his tools. And, despite the best intentions in all three areas, making a

These 16th-century English earthenware floor tiles display a barrel motif.

barrel would still have been difficult. Every split, cut and shave of the barrel's wooden components had to be made solely by hand tools; accuracy would have been difficult, especially with the harder *Quercus robur*, or English oak, and the equal lack of quality control within the steel and tool-manufacturing operations of the era. What with small shaping and cutting errors due to cruder tools, combined with more difficult conditions of usage (rolling on cobbled streets, lack of consistent climate control and long years of use), barrel leakage was definitely more of a problem than it is today. This would have necessitated coopers being continually available to repair the barrels as they arrived on ships, were loaded on and off wagons and were used in the wine-storage warehouses.

The earliest coopers were also probably a fairly independent lot. However, as the demand for their services increased, and rising incomes brought the taxman to their door, they, like most of the earliest craftsmen, soon realized that they would have more power to control their businesses if they combined into fraternal organizations than if they stood alone. These associations allowed master coopers to maintain the independence of their own shops, as opposed to creating larger, multi-employee cooperages, while also providing political and administrative mechanisms to fend off bureaucratic tax demands, control unscrupulous coopers, encourage minimal standards of barrel capacity and quality for their customers, train apprentices properly and provide for coopers and their families in old age and in the event of accidents or death.[51]

Within the world of medieval trades, coopers were among the defined crafts conducting business in London, Britain and mainland Europe. By 1196, guild rolls in Leicester listed, in addition to the coopers, the following trades: bakers, butchers, carpenters, cooks, dyers, farriers (blacksmiths), goldsmiths, hosiers (stocking and sock makers), leather workers, leechers (early physicians), masons, mercers (dry-goods merchants), millers, ostlers (horse keepers, particularly at an inn), parchment makers, plumbers, porters, potters, preachers, saddlers, shearers, shoemakers, soapmakers, tailors, tanners, turners (makers of rounded wooden objects), watermen, weavers and wool combers.[52] Trade lists in London and Bristol would have included vintners, the merchants who purchased and sold wine.[53] While almost all these crafts still exist today in some form or another, thankfully the physician 'leechers' have found other methods of healing! As a comparison, the guild rolls of 1363 from Nuremberg, Germany, added only clothiers, locksmiths, coach-builders, hatters, toolmakers, plasterers, glaziers and painters[54] – trades which were probably also encompassed by some of the Leicester guilds.

Despite being one of only about 40 designated crafts, coopers were still a relatively minor group and remained so throughout history. Records indicate that around 1377 there were only twenty coopers in London, out of a city population of 35,000. As the population grew to 60,000 in 1540, about 50 coopers were listed on the tax rolls. By 1640, just at the start of the heyday of sail, still only some 75 coopers were working, out of a population of 300,000.[55]

Over the following three centuries there was an explosive demand for barrels to contain beer, rum, whale oil, herring, sugar and numerous other commodities, which required the training and engagement of thousands of coopers – but they were still a minority trade group. Since goods such as whale oil have been completely phased out, and the other products have changed their packaging needs, demand for coopers has been reduced to currently no more than 100 coopers in all of the United Kingdom, with similarly small numbers, relatively speaking, in the U.S. and Europe.

## Coopers in Mainland Europe

As we've discussed, it appears fairly evident that it was Celtic crafts-men who first made barrels. We've also seen how they had gradually become part of the everyday scene by the Middle Ages, especially in northern Europe with its plentiful supply of mixed-species trees, and along the way how barrels had become more plentiful as the

A drawing by Jean Frédéric Wentzel, c. 1847, with details
of the hand tools utilized by the cooper.

Until the mid-20th century, barges and boats, and the rivers and
canals they plied, were the main vehicles for transporting barrels.
These are wine barrels in transit at the Paris entrepôt of Bercy.

Romans shipped wine to all parts of Europe and later introduced
wine growing into northern Europe.

Records from Germany indicate that vineyards were widespread
in the Würzburg region by AD 779.[56] Würzburg is in central Germany,
roughly 100 km southeast of Frankfurt and only about 200 km north
of Heuneburg, the ancient Celtic site. As such vineyards became
fully developed, with bountiful wood resources nearby and ready
markets both domestically and abroad, the cooper's skills were much
in demand for the barrels and tanks utilized in many of the winemaking processes,[57] and also extended into making barrels for other
products. With aspirations similar to those of their brother coopers in France and Britain, German coopers also wanted to control
their products and protect their markets. To do so, they too formed
associations. In Würzburg, they were declared professionals in 1373,[58]
and the tradesman list in Nuremberg included 34 coopers doing
business in the city.[59]

To a lesser extent, coopers and cooperages also existed in Eastern
Europe and down on the Iberian Peninsula – wherever the need

arose for wine barrels or everyday buckets, tubs and tanks during the medieval period. Coopers in those regions were all feeling the similar growing pains of an expanding industry – cyclic economic periods, finicky customers, inconsistent supplies and shifting alliances.

# SIX

# Parallels: Wooden Barrels and Wooden Boats

Day after day, day after day,
We stuck, nor breath nor motion;
As idle as a painted ship
Upon a painted ocean.

Water, water, everywhere,
And all the boards did shrink;
Water, water, everywhere,
Nor any drop to drink.

Samuel Taylor Coleridge,
'The Rime of the Ancient Mariner', 1798

 We have already seen how, with the advent of sailing ships capable of encircling the globe, barrels came to play a vital role in storing both the ship's supplies and its traded goods, and that the parallels and close relationship between boat- and barrel development are well worth exploring. This chapter focuses on telling this barrel/ship story in greater detail.

One ship's tale involves the very famous *Mayflower*, an English ship of 180 tons – slightly larger than the *Asterix*. After several years employed as a general cargo ship, which included transporting English woollens to France and returning with barrels of wine, in 1620 it was chartered to carry Pilgrims, separatists from the Church of England, to the New World.[1] Sailing was set for July of that year, but was delayed due to the fact that the original ship's cooper/carpenter decided not to make the journey. A replacement needed to be found as the maritime law of the times forbade a ship to sail without a cooper on board. Finally, one John Alden was hired. After further delays and rough Atlantic seas, the *Mayflower* arrived off Plymouth, North America, in late November, at the start of a difficult winter which saw the death of a number of its passengers. Mr Alden was one of the survivors and later went on to be the assistant governor of the Plymouth Colony.[2]

To serve as both ship's cooper and carpenter was not uncommon, as the woodworking skills and tools were quite similar. In the days when vessels plying the oceans for whaling, merchant and naval purposes were wooden, it was essential for each ship to have one or more men who could repair spars, oars, small boats and the vast array of wooden seagoing equipment, as well as assemble and repair the barrels which stored the supplies – such as drinking water, salted meats and grains – and trade items like whale oil.

Comparable to the port pipes sitting in the barcos rabelos, the story of the Mayflower and Mr Alden illustrates the close connection between wooden barrels and wooden ships that we started to look at earlier in the book. As we saw, the evolutionary development of both barrels and ships began far in the past and shared many similarities in terms of raw materials, tools and techniques – and ultimately the functions which they served in advancing civilizations. As such, examining wooden boats, like studying the use of amphorae, is helpful in understanding the role and evolution of barrels throughout history.

We also saw that, historically, the development of wooden barrels, starting with wooden buckets, evolved within a time frame roughly similar to that of boat building – if one could hollow out a log, one could make a boat or bucket. This was especially true in the lands around the Mediterranean Sea. At least a couple of thousand years ago, as cultures developed the stone, bronze, copper and eventually iron tools needed to cut and shape wood, people also developed a comprehensive understanding of the various tree species and which ones were best suited for specific purposes. Subsequently, as the metals used to work the wood, and the technology to fasten the pieces together, became more sophisticated, both barrels and boats could be constructed with greater efficiency – ultimately for greater numbers of products or longer and longer voyages.

Barrels and ships were especially close partners during their golden ages: the fifteenth to nineteenth centuries. During that period, William Bryant Logan noted in his excellent book about oak trees and their contributions to both barrels and wooden ships that 'the power and reach of western and northern European cultures grew by powers of ten . . . journeys went from hundreds of miles . . . to crisscrossing the globe'.[3] Those ships carried many barrels, but of course in recent times, boats' and barrels' paths have diverged. Metal containers and tanks, refrigerators, freezers, cans, stainless steel tubs and plastic containers now hold most of the stores and supplies onboard ships, yachts and boats, while boat manufacture has veered from wood towards steel, fibreglass and aluminium. And while finer wines and whiskies continue to be aged in wooden barrels, the majority of those products are stored in tanks of stainless steel, fibreglass, coated concrete or plastic.

## Constructing Boats and Barrels

A wooden boat is designed to keep fresh or salt water out, while barrels are also designed to keep the elements out and their contents safely inside. One of the methods of fabricating wooden boats – placing strips of wood tightly together side by side and caulking the gaps as necessary – is similar to that used in building barrels. The use of many strips of wood, the types of wood used and some of the barrel-construction methods employed, such as bending wood with heat, are comparable to those used for wooden boats.

As I write this book, I look out on the small but busy seaport of Nelson, New Zealand. Through this port, many of the products produced in the Nelson and Marlborough regions leave for export and other products arrive for consumption. (Among other indigenous items are containers full of cases of Marlborough Sauvignon Blanc, although unfortunately most of this is made in large, stainless steel tanks and not wooden barrels!) Bulk-carrying ships come to pick up the pine logs that are heavily cultivated throughout the South Island, and container ships arrive to receive refrigerated containers of apples, pears, frozen fish, lamb, mussels and cases of wine. It was not too many years ago that many of these products would have been shipped in wooden barrels.

Within the port, fishing boats and pleasure craft have their own separate, sheltered marina, with several shipyards hugging the shoreline. One day, while walking by one of these repair yards, I noticed an older wooden fishing boat set out at the back of the lot. It was basically just a deteriorating hull, faded and with its red paint peeling away. Between the strakes, or planks of wood which comprised its hull, there were gaps which indicated that the timber had dried out as the boat languished in this storage area. I have seen many barrels do the same. What did surprise me was seeing sections of oakum, or tarred cotton string, hanging out of some of the gaps. It had been used to caulk the joints – that is, to fill in the small gaps between the pieces of wood. A little research showed that this is a common procedure in constructing wooden boats. It is also a common, although less than satisfactory, repair method for wooden barrels and tanks. On a boat, the exterior water pressure forces the oakum in, while on a barrel or tank, the pressure from the interior liquid pushes the oakum outwards. With a boat, the oakum would also provide some measure of water resistance while the wood was swelling when the boat was initially placed into the water. This type of boat construction, where the planks butt up against each other, flush, and are fastened to the framework by screws or dowels, is termed 'carvel'. It was a common construction method for ships built for the Dutch, English, Portuguese and Spanish navies during the Middle Ages,[4] and, as observed in this fishing boat, is still in occasional use today. The carvel-type construction is directly analogous to the way the staves of barrels abut: side by side, parallel to each other. And the strakes are heated so that they can be bent to conform to the curvature of the boat's hull lines, just as barrel staves are also heated in order to be bent.

Another type of boat construction with similarities to barrel construction is known as lapstrake. In lapstrake, the outer planks overlap one another and are fastened together either with fibres, bits of saplings, treenails (wooden dowels) or, in more recent times, copper nails or stainless steel screws. Excavations of early Viking longships revealed this technique.[5] The similarity to barrel construction is not specifically in the lapstrake form of construction, but in the methods of securing it. For barrels, saplings were made into hoops to contain the staves, fibres hold the sapling hoops together and wooden dowels are used to align and secure the headboards together.

However, some construction techniques have not been equally successful in both wood-crafting spheres. Lamination is a newer boat-building technique that came too late for general barrel construction and is considered unacceptable for today's wine or whiskey barrels. Laminating thin strips of wood together has its origins in the Egyptian mummy cases and was adapted by Chinese craftsmen for lacquered boxes and furniture.[6] Now commonly used in a boat-construction procedure called cold-moulding, the layers are built up over an interior framework. Utilizing today's excellent glues and resins, a strong, watertight hull can be formed without the need for oakum or other types of caulking.

Lamination has not been successful for barrels because early glues, made from gelatine derived from animal hooves and wastes, either negatively impacted the product within the barrel or were not strong enough for barrel construction. While a boat would normally have an internal framework with which to support the laminations, a barrel would become too heavy with an interior framework, but, without one, would not have enough strength to withstand the forces of moving and handling. Neither would an exterior framework be suitable, as it would defeat the barrel's design for mobility.

In current times, the relatively limited supplies of the clear, all-heart oak used to craft wine and whiskey barrels have become more expensive, thus increasing barrel prices. Attempts to produce less expensive barrels through the lamination process, by incorporating smaller and thinner pieces of the best wood on the inside of the barrel and using lesser grades for the bulk of the exterior wood, have failed. Again, even with advances in fabricating inert synthetic glues and resins, winemakers and distillers have serious concerns about their potential interaction with the wines and whiskeys. And even if proved non-reactive, the glues would not allow the essential oxygen interaction with the liquid which makes the wooden barrel so unique. For similar reasons, plywood barrels have not been acceptable.

## Tools

Today, more and more of the hand tools utilized by boatbuilders and coopers incorporate levers, triggers and handles, plus the dials of motorized apparatus and the touch screens of computers. The wood is only handled when placing it in the machine or positioning it in the boat or barrel; otherwise, almost every operation is done by a machine, either hand-held or fixed in place, and now often computerized. But this change from physically utilizing hand tools to crafting both boat and barrel totally by machine is relatively recent, and has occurred only within the past 150 years.

Until the Industrial Revolution, the trees used to build both boats and barrels were felled by axes and saws. The hand tools utilized by the nineteenth-century cooper had seen little change from those developed and used by the Celts.[7] Logs were split by wedges or froes (wedge-shaped tools with a handle at right angles to the blade), driven in by heavy mauls or sawed into planks with saw pits. They were trimmed and sized by broadaxes and adzes. Thereafter a myriad of augers and bits, hand knives, chisels, gouges, files, shaves, drills and planes for detailed shaping, jointing, rabbeting, grooving, scraping, planing and trimming would be employed to bring each piece to its final configuration. Measuring devices were part of the cooper's and shipwright's tool chest as well, with dividers and angle jigs for barrels and squares, bevel gauges and various early tape measures for boat building.[8]

## Wood

If a nation has ever gone to war to protect its oak forests, it has not been well documented. But most likely every other type of conflict – physical and legal – has occurred as landowners, timber companies, cooperage and boat-building industries, governmental agencies and navies struggled to ensure a continuing supply of oak, the primary material for both barrel and ship construction.

As early as 1543 politicians in England realized they were overcutting their oak forests. They subsequently passed laws to control the use of oak by restricting the export of large casks.[9] By the midnineteenth century, after several hundred years of frantically building more and larger merchant and naval ships, the quality timber – 'oak for knees and planking, tall, true pine for masts', penned Andrew Lambert – within European forests was all but depleted.[10] Jean-Baptiste

Colbert, who served as Louis xiv's Controller of Finance and Secretary of the Navy, issued strict edicts to protect French forests in order to ensure a supply of timber for their ships, especially their naval ships. He set aside certain forests, such as the famed Tronçais stand of oaks. The French government would come to manage this tract, and still does, although it is now a significant source for wine barrel and furniture oak. And he decreed that timber could not be harvested within 76 km of the sea or 29 km from inland waterways, to minimize bootlegged timber. Additionally he prevented timber trusts and businesses from collaborating to inflate timber prices. Ironically he is also famous for designing a 74-gun ship which would become the large (but standard of the day) man-of-war sailing ship.[11]

War, disruptions to transportation systems and rising material costs often temporarily impacted coopers' and boatbuilders' sources of wood and other raw materials. As the English forests were depleted, English coopers and shipbuilders turned to oak from northern Poland and eastern Russia, largely shipped out of the Baltic port of Memel,[12] as noted previously. While this timber was the same oak species, *Quercus robur*, as that found in England and France, it became known as Memel oak. During the First World War this source was again interrupted, and the English coopers had to make do with their dwindling supply of domestic oaks, as well as some obtained from America.

According to Maynard Bray, a wooden boat builder, the phrase 'use pine where you can and oak where you must' was an adage from early American ship-building days.[13] The oak could be bent and was strong at the critical connections with the ship's frame, such as the ends and joints along the keel, but pine, cheap and more widely available, could be used on the long, relatively straight sections, especially above the waterline.[14] Barrels made of pine have been used primarily for slack barrels for dry goods, since it neither bends well nor is particularly waterproof. Greek and Cypriot retsina wine is made in pine barrels, but these are coated on the inside to prevent leakage, which tends to give the wine its resin-like taste and thus its name.

## Metal Fittings

This is where the construction of boats and barrels starts to diverge: the main use of metal in barrels is primarily for the hoops and for the rivets which secure them. Additionally brass or stainless steel spigots are occasionally used in wine barrels to draw the wine. And when beer was delivered in barrels, a specialized metal bung fitting kept the pressure in the barrel and allowed the attachment of a tap. In recent times, especially for wine barrels, in which the staves swell when the barrel is full but shrink when it is empty and dried out, and particularly in the drier climates of California, Australia and Chile, the hoops are secured with small L- or T-shaped nails, unimaginatively termed 'hoop nails'.

As metal fabrication became common, and the range of usable metals diverse, all sorts of nautical fittings – winches, pulleys, braces and eventually wire rope or steel cable – were developed for wooden sailing and powerboats. The last great sailing ships, stout German vessels designed to round Cape Horn for the lime trade, had both steel and wooden hulls and replaced much of the hemp rope in the rigging with steel wire. Just as the near-universal use of wooden barrels gave way to containers of steel, plastic and cardboard, the large wooden-hulled ships disappeared from the sea within a relatively short period of time. Some smaller craft continue to be built from wood, but overwhelmingly steel, lightweight aluminium, fibreglass and carbon fibre are used to construct everything from dinghies to huge supertankers and world-circling racing yachts.

## Barrels on Board

While wooden ships and wooden barrels evolved within roughly the same time frame, utilizing much of the same toolkit and technology, is there any evidence to suggest that the ships were designed specifically to contain barrels? The V shape of a ship's inner hull accommodated barrels and, as we've seen, was one of the reasons for the shift from amphorae to barrels as the Romans transported wine from mainland Europe to Britain.[15] Pictures, reliefs and archaeological digs regarding the ships of 2,000 years ago show that barrels were carried aboard many different kinds of ships. But while barrels may have been the primary container, just as today's generic truck vans are designed to carry many types of freight, so it appears that most sailing ships were not purpose-built for any particular kind of container or cargo.

Following this line of inquiry further, did shipbuilders incorporate special structures within ships' holds for the barrels? And, related to this, how did sailors keep the barrels in place during rough voyages? Some answers to these questions were provided by a visit to the wooden sailing ship, the Edwin Fox, which is preserved within a covered dry dock at Picton, New Zealand. Unrestored, the Edwin Fox now consists only of the hull and a portion of the single interior deck. With the entire upper weather deck removed, and deterioration in some portions of the hull, the amazing details of the wooden framework, plus the painstaking craftsmanship that built it, are clearly exposed. In this condition the Edwin Fox provides an ideal example to see how barrels, and other cargo, could be stowed in wooden ships.

The Edwin Fox was constructed in Calcutta for British owners in 1853.[16] Its framing is of teak and saul, both hardwoods. Like many

Ships and barges carried various commodities in barrels.
These were being moved through the Dardanelles.

merchant ships of its day, it travelled throughout the globe – from Britain to Australia and from China to Cuba – carrying a variety of cargo, including cotton, rice and tea, and also convicts, who were kept in small enclosures. Products were enclosed in a multitude of packaging: cloth and fibre bags and wooden barrels, cases, crates, boxes and spools. During the planning and building of a general-merchandise ship such as the Edwin Fox, neither the ship's designers nor builders, nor subsequently the ship's captain or boatswain (bo'sun), its loading supervisor, would have had any foreknowledge of the specific types or quantities of cargo that the ship was likely to receive over its many years of service. Thus the hold was kept relatively unencumbered of framing or bulkheads in order to maximize each load.

Without any special framing built in to hold the barrels, the essential concern was to secure the barrels, or any of the cargo, against movement during the voyage. A loose barrel on a rolling ship would have been as dangerous as a loose cannonball. To accomplish this, the ship's bo'sun and his crew would have utilized all kinds of dunnage – the general term for pieces of repair wood, lines, sails, other bits of ship's equipment or even unbreakable cargo – to pack in and around each barrel and other items of trade. The dunnage would also have included firewood for the ship's stove(s), dimensional planks and boards for repairs, staves and barrel heading and small, wedge-shaped chocks and stops to secure the barrels further.[17]

On visiting the ship, it piqued my curiosity to figure out just how many barrels could be placed within the hold, with or without any special bracing. I wanted to know whether the total weight of the barrels or the capacity of the hold would be the limiting factor. For a standard-size 225-litre barrel full of liquid, the total weight is 263 kg. Start placing in a ship a quantity of these barrels for the ship's water and food supply, plus any number for the cargo, and very soon, or so I assumed, the weight would become a significant dynamic.

However, a short description of the hull of the Edwin Fox will disperse any concern about the weight of the barrels and their impact on the ship's integrity. The ship was about 43 metres long and 11 metres wide – larger than the Asterix, the Mayflower or Christopher

Columbus's ships, but still tiny compared to the steel-hulled, general-merchandise ships of today. The Edwin Fox's hull consisted of four layers. The internal frame was built of huge, 30-cm square timbers, densely set at 30 centimetres apart. Attached to the outside of this framing was 16-centimetre-thick teak planking butted tightly together in the carvel fashion, edge to edge like the staves of a barrel. Much of the planking at and below the water line was placed diagonally, which would have added significant strength to the hull. On the outside of this planking the Edwin Fox had copper plate, nailed on to provide additional waterproofing and protection against worm borers. On the interior side of the framing, planking of least 10 centimetres thick was attached. The thickness of the entire hull was somewhere near 60 centimetres. Thus a load of heavy, full barrels was absolutely no contest pitted against a hull of this massive strength.

Looking at a cross-section of the Edwin Fox's hull, from the keel the sides of the hull curved up, out and then up again, in a kind of sinuous v shape. The ship's sides were locked together at the top with the horizontal deck structures. The weight of the cargo or barrels would press the sides out, while the seawater readily resisted those forces and pushed the sides inward. Thus the weight of the barrels, while not insignificant, was easily countered by the strength of this thick hull and the press of water on the outside of it. According to certain descriptions of the ship, it actually needed some 400 metric tons of cargo and/or ballast for proper sailing – much more weight than a complete load (possibly some 500–750) of full barrels!

## Loading and Unloading Barrels

Moving a several-hundred-kilogram barrel around a gyrating ship on the open ocean in order to get to new provisions or water, or to fill it with whale oil, was a perilous task. It called for skilled hands and securing procedures while the barrel was being shifted.

Getting the barrels on and off the ships and in and out of the holds was also an issue. Deep-water piers and wharfs were almost non-existent, even during the height of the sail-shipping era.[18] Few

Examples of some of the rigging used to hoist barrels on and off ships and boats.

city harbours had stone quays. Where a quay existed, a ship could tie up close, but it then needed a ramp to roll the barrels on and off. More often, ships would be anchored out in the river or harbour. Lighters ferried the barrels and other supplies between the wharf or beach and the ships. Additionally barrels were often rafted – tied together – and floated to or from the ship.

To transfer cargo to or from the ship via a lighter or the wharf, a lower yardarm of the mast, the lowest horizontal beam from which the sails hung, would be employed, acting as the boom of a crane. With block and tackle cargo could be raised or lowered. Sailors would control the lines, directly pulling on them or attaching them to a man-powered winch for heavier items.[19] This yardarm, with its double pivot connection to the mast, was capable of being swung out over the side of the ship. When transferring cargo from a lighter to a ship, for example, barrels would be hauled up, either by grabbing the chime (the ends of the staves) with hooks, by tying them up with line utilizing special knots or by placing them in cargo nets. Once hauled up and over the ship's railing, the yardarm would be rotated back over the gunwale to set the barrels down on the deck or to lower them through a hatch into the ship's hold.

## The Ship's Cooper

As we have seen, towards the end of the Roman era, the wine trade boosted the numbers of barrels transported at sea. And as the push to expand from the Mediterranean and Europe to the Americas and Asia increased, barrels accompanied all voyages. Twede notes that the wooden barrel was extremely important to the imperial expansion of British explorers, and this was equally true for the Dutch, Spanish, Portuguese and Italian voyagers.[20] (Moving up to the fourteenth and fifteenth centuries, the next big increase in the numbers of barrels on board a ship was for the use of whale oil.)

On board sailing ships, a cooper such as John Alden, and the sailmaker and cooks, did not stand watch because they had normal day duties. But, like everyone else on a sailing ship, when the vessel was tacking to windward or wearing (turning from the wind), regardless of whether it was night or day, they were expected to be at their sail- or line-handling station.[21]

A good cooper, like a good cook, could help to make a successful voyage. In 1722 a cooper on board a ship under the command of the Dutch admiral, Jacob Roggeveen, the *Africaansche Galey*, kept his casks airtight by tightening the hoops frequently and making sure they were in good repair.[22] As a result that ship enjoyed dry food, whereas the other ships of the fleet suffered from rotten food. One wonders why the other captains did not demand the same of their coopers, although the understanding of the principles of food preservation and spoilage was still in its infancy.

## Barrels and Captain Cook

British explorer and navigator extraordinaire Captain James Cook (1728–1779) was extremely industrious and tried all kinds of things to improve shipboard life. Like Admiral Roggeveen, he also had his coopers periodically check the dryness of the commodities within the barrels, especially the gunpowder.[23]

Wooden barrels were excellent for storing dried foods, but were less effective for fresh fruits and vegetables, especially on extended three- to four-month sea voyages. And without these fruits and vegetables, and the vitamins they provided, scurvy was a significant problem for sixteenth- to nineteenth-century long-distance sailors. Far more British sailors of that period died of scurvy and other nutrient-deficient diseases than from wounds inflicted by gunfire or cannon shot during the naval battles.[24]

By incapacitating the crew, scurvy reduced the efficiency of the ship; therefore, most captains were known to be, to some degree, concerned about its effects. Captain Cook is particularly notable, not only for his three, multi-year Pacific Ocean voyages of exploration, but also for his success in minimizing scurvy within the crews of his ships *Endeavour* and *Resolution*. For it was during those voyages that Cook actively loaded aboard, in barrels, as many foods as he could which supplied the nutrients needed to prevent scurvy. He detailed these specific efforts in his extensive diaries, and those efforts and successes have been substantiated by the diaries and letters of other members of his crews. In his diaries, he also described the use of barrels and cooperage and how the barrels played a role in his attempts to keep his men clean and maximize their health and well-being on those extended expeditions. Cook even built his defensive redoubt at Tahiti by filling barrels with dirt and lining them up as breastworks.[25]

Cook was born in 1728 in North Yorkshire, England, and died on 14 February 1779 in a deadly encounter with native Hawaiians in Kealakekua Bay, Hawaii. At fourteen, he left school and trained as a shopkeeper, then apprenticed with a provisioner of ships sailing the North and Baltic Seas out of the British port of Whitby and later served as a sailor on ships taking coal to London and on trading ships traversing those same seas. The knowledge of stocking a ship and navigation gleaned from these employments enabled him to progress from sailor to naval officer during assignments to Newfoundland, and ultimately he was placed in charge of the difficult and important explorative Pacific journeys.[26]

Over a period of eleven years, from 1768 to 1779, Captain Cook 'discovered' – at least for the Western world's knowledge base – and charted many of the Pacific Islands by transiting the Pacific from South America to Australia and from Antarctica to the Arctic Ocean.[27] Historians attribute Cook's success to a number of reasons: he was a superior leader and well respected by most of his men; he was an exceptional navigator, and despite the rudimentary navigational tools of the day (the precision clocks required to measure longitude and distance accurately were just being introduced), he made detailed and accurate charts; the naturalist and painters who accompanied his voyages were afforded numerous opportunities for interaction with the native Pacific islanders, as well as for viewing and recording the flora and fauna of those islands; and he discovered and explored many islands (noting their resources and potential for trade), as well as a potential Northwest Passage sea route and the limits of Antarctica (to the extent that he confirmed that there was no continental paradise at the southern end of the globe). So perhaps it is unsurprising, especially given his shopkeeping background, that his attention to detail even extended to the wooden barrels that held his water and supplies; he knew that these were critical to the health of his crew and the success of his journeys.[28]

Although Cook is noted for his conscientious and continuous efforts to find foods that would reduce or eliminate scurvy, his new ideas sometimes backfired. At one point he encouraged the men to eat walrus flesh, which was apparently an experience somewhere between eating rotten fish and chewing on a sweaty horse blanket. Even the saltiest sailors gagged. On this particular voyage, a midshipman by the name of Trevenen wrote in his diary that 'Captain Cook here speaks entirely from his own taste which was, surely, the coarsest that ever mortal was endowed with.'[29]

The sailors of Cook's time were an obstinate bunch and creatures of habit when it came to trying new foods. On occasion Cook used some force to get them to eat scurvy-preventing foods, but he mainly relied on example. After he first served sauerkraut in the officers' mess, it became seen as a 'luxury' that the sailors then also desired.[30] One of Cook's biographers, Nicholas Thomas, noted the foods that Cook made the effort to acquire for his officers and men in order to combat scurvy: 'malt, sauerkraut, salted cabbage, soups, lemon and orange concentrate, carrot marmalade and the "inspissated juice of wort" – a sort of base for beer [which would have contained vitamin B]'.[31] All of these provisions would have been carried in wooden barrels.

Rum was part of the British sailor's rations, but in an effort to run a sober ship, Cook allowed it to be served up only at special times, such as Saturday nights and Christmas.[32] He also tampered with its contents – to minimize its alcoholic effects as well as prevent scurvy – by mixing the rum with three parts water and a bit of lime or lemon juice.[33] And all those who enjoy a Piña Colada cocktail can thank Captain Cook, who also added coconut milk to his sailors' rum drinks – again to act against scurvy.

Cook's efforts at preventing scurvy provided measurable results. Here, from his printed diaries (1777), is his account of meeting up in New Zealand with the sister ship Adventure on his second voyage:

On the 29th [July 1773] I [went] on board the Adventure [the sister ship sailing with Cook on this second voyage and meeting him at Ship's Cove in Queen Charlotte Sound, New Zealand, near Picton] to inquire into the state of her crew, having heard that they were sickly; and this I now found was but too true. Her cook was dead, and about twenty of her best men were down in the scurvy and flux. At this time we had only three men on the sick list, and only one of them attacked with the scurvy. Several more, however, began to shew symptoms of it, and were, accordingly, put upon the wort, marmalade of carrots, rob of lemons and oranges.

I know not how to account for the scurvy raging more in the one ship than the other; unless it was owing to the crew of the Adventure being more scorbutic when they arrived in New Zealand than we were, and to their eating few or no vegetables while they lay in Queen Charlotte's Sound, partly for want of knowing the right sorts, and partly because it was a new diet, which alone was sufficient for seamen to reject it. To introduce any new article of food among seamen, let it be ever so much for their good, requires both the example and authority of a commander; without both of which, it will be dropt before the people are sensible of the benefits resulting from it. Were it necessary, I could name fifty instances in support of this remark. Many of my people, officers as well as seamen, at first disliked celery, scurvy-grass, &c. being boiled in the peas and wheat; and some refused to eat it. But, as this had no effect on my conduct, this obstinate kind of prejudice by little and little wore off; they began to like it as well as the others; and now, I believe, there was hardly a man in the ship that did not attribute our being so free from the scurvy, to the beer and vegetables we made use of at New Zealand. After this, I seldom found it necessary to order any of my people to gather vegetables, whenever we came where any were to be got, and if scarce, happy was he who could lay hold on them first.[34]

Cook's voyages traversed some of the world's roughest seas – around Cape Horn at the tip of South America and the cold, southern waters around Antarctica. And without the long-range weather forecasts upon which modern mariners depend, he inevitably encountered some horrific storms. With the high seas in these storms, coupled with sailing in equatorial regions, wetness and humidity on the ship was a constant problem for all his stores. In an attempt to combat the humidity Cook had his carpenter/cooper (who was James Wallis on the *Resolution*) line some casks with tin foil, which proved helpful in keeping dampness from his flour, malt, oatmeal, groats, peas and bread,[35] and also periodically check the dryness of the commodities within the barrels.[36]

On another voyage Cook discovered mouldy bread in a number of casks. The cause turned out to be barrels made of green, inadequately seasoned wood.[37] Cooperages manufacturing barrels for these long-distance voyages normally utilized only dry staves. Placed in a barrel, green staves could add moisture or continue to shrink, causing gaps or openings into which air or water can enter, spoiling the contained product. Additionally, iron hoops, as opposed to fibre or sapling hoops, were generally used on barrels for extended journeys, again to ensure that the staves could be securely tightened and preventing any such movement. It is not clear where these barrels originated; whether they were made by Cook's cooper, supplied by a dishonest ship's food purveyor, or manufactured by an unscupulous cooperage.

On Cook's second voyage, he left the Cape of Good Hope, South Africa, in November 1772, sailed south and east, pushing into the Antarctic seas, and eventually arrived in Dusky Sound in the Fiordlands of New Zealand on 26 March 1773, four months later.[38] He would have been able to get some fresh water from rain and snow, catching it in the sails. He may have stopped at one of the desolate southern islands, but water would have been all he could obtain in terms of supplies. We might assume that he went for two months at a time between stops to replenish his water supply. For a two-month stint, Cook and his crew would probably have needed at least 96 barrels just for water (90 men x 60 days x 4 litres per day[39] = 21,600 litres divided by a 225-litre/barrel). He also had to have several hundred barrels for basic food supplies to last a year or more, and probably 100 more as he expected to be out for several years. All these barrels were placed on a ship just 34 metres long. The ships of that day carried many barrels, squirrelled away in every conceivable and available space. Keeping them in good shape, and moving them periodically to get to the supplies they held, was for Cook and his crew a constant but successful effort.

## Barrels and Whaling

Wooden barrels were also critical to whalers: besides holding their water and supplies, they provided the containers for their whale oil. The most valuable whale, the sperm whale, could yield up to 80 barrels of oil. Whale barrels would have been in the 100–300-litre range, tending towards these smaller sizes in order to facilitate moving them around on a rolling and pitching ship.

On board a whaling ship, to save space on the outward journey for the ship's food and water supplies, the cooper would 'pre-package' staves and heads, storing them within other barrels or bundling them into shooks. When a whale was captured, the whale meat would be melted down for oil. This was termed 'trying out' and was accomplished in try-pots – huge iron cauldrons (often with two flat sides) set in ceramic bricks on the ship's deck and fired with additional whale meat. In preparation for trying out the ship's cooper would reassemble the barrels to be filled with the rendered oil. As the provisions – food and water – were used up, the barrels which contained those items would be refilled with the oil. When casks were not needed, the cooper would break them down. This was termed 'shaking' the casks.[40]

John Hector St John de Crèvecoeur was a French immigrant who fought in America's French and Indian War and later went on to become an American citizen, establishing a farm in New York State's Orange County. As a man of some means, and an astute observer of American life in the late 1700s, Crèvecoeur travelled throughout the U.S.'s eastern seaboard and countryside, providing a vivid written description of early American life. Included in his travels were trips to Martha's Vineyard and Nantucket, islands just south of Cape Cod, to report on the whaling industry. At that time Nantucket was largely a Quaker village and an enormously successful whaling port, with a fleet of up to 300 whaling ships. Crèvecoeur put these observations in a book, *Letters from an American Farmer*, in which he describes part of Nantucket's successful infrastructure:

> At schools they [the Quaker boys] learn to read, and to write a good hand, until they are twelve years old; they are then in general put apprentices to the cooper's trade, which is the second essential branch of business followed here; at fourteen they are sent to sea, where in their leisure hours their companions teach them the art of navigation, which they have an opportunity of practising on the spot . . . Then they go gradually through every station of rowers, steersmen, and harpooners; thus they learn to attack, to pur-sue [sic], to overtake, to cut, to dress their huge game: and after having performed several such voyages, and perfected themselves in this business, they are fit either for the counting house or the chase.[41]

Interestingly, throughout the remainder of his book he barely mentions the wooden barrels or the coopers who made, cared for and repaired those barrels. This omission is somewhat puzzling as he notes that the coopering of barrels was an extremely important – he even utilizes the term 'essential' – aspect of the whaling business.

On the other hand, perhaps the omission should not surprise us. During the period in which he writes, barrels were so ubiquitous that they were no doubt taken for granted. Perhaps because they were so common in both America and his French homeland, for whose people he was primarily writing, there was little need for further descriptions.

Around the world, whenever a whale washed up on an inhabited beach, most likely the locals took from it what they could – food, oil, baleen or bone. For the earliest peoples, the oil – for light and heating – might be kept in skins, ceramic jars, wooden buckets, shells, gourds or stone pots. Even early fishermen who hunted whales from small boats in the open oceans only caught a few, and so there was little need for a designated container to hold the whale oil on board the boat or transport it once ashore. But as whaling ships increased in size and larger whales were caught, often in the far corners of the world, some convenient container to store and transport the oil was necessary. Wooden barrels fulfilled that requirement nicely.

Whaling, where the ships carried wooden barrels on board to store and bring home the oil, may have started as early as AD 1000. But many years passed before the peak of whaling started in the seventeenth century. Only by then had ships become large enough to travel the oceans and process the largest whales. An additional impetus came from growing demand for the oil: the citizens of Europe's and America's cities wanted light in the form of whale-oil candles, which burned more cleanly than ones made from other animal fats.[42] Further, the exponential development of precision instruments during that era required the very finest oil for lubrication: that oil came from the huge sperm whales. But by the mid-nineteenth century, evidence of over-harvesting was apparent. Ships were required to go further afield, around the Pacific and towards the poles, spending longer periods at sea for fewer whales.[43]

When a whaling ship was at sea, as Australian maritime historian Granville Allen Mawer notes, the 'stowage of the oil was the responsibility of the cooper and leakage was his enemy'.[44] One whaling-ship cooper was a certain William H. Chappell. In his diary, written between 1852 and 1855 while sailing on the whaling ship *Saratoga*, based in

Martha's Vineyard, he notes that in preparing the cooperage for the whale oil, he set up the staves of a cask, but in one instance needed to construct new heads for that cask. Apparently the original heads had been left in port 'through negligence'.[45]

Once a barrel has had any food or liquid in it, like any other container, it must be cleaned. And because of wood's porosity, and depending on what was in the barrel, some sort of preservation was usually necessary. On whaling ships, salt water was a common preservative (and salt water in casks also provided ballast for the ship). This water would be changed every few weeks, or every few days in warmer climates. Nowadays, sulphur is burned in wine barrels for its antimicrobial effects, and in those days too it was used as a preservative, especially in ships sailing from regions where wine barrels were in common use. Also, when the shooks were set up, they would need a period of swelling to get the staves tight, and a couple of gallons of hot whale oil were often used. This boiling oil also doubled as a sanitizing agent, killing any offensive microbes hiding in the wood.

As seen in the cross-section overleaf, casks of several sizes were stowed throughout a whaling ship. The whaling bark was about 40 metres long and 8 metres wide. Placed on two decks, the larger casks would be below the try-pots, while smaller ones are scattered about wherever space permitted. Because it would be difficult and dangerous to move the large casks while at sea, they were positioned beneath the try-pots for ease when filling.

Prior to 1824 whale oil was measured in 'barrels' of 31.5 gallons (119.2 litres) and 'tuns' of 252 gallons (953.9 litres) or eight barrels. A whaling ship's tonnage was the number of 'tuns' it could stow – a continuation of that measurement instigated by the Romans. Then, in 1824, the British abandoned the term 'tun' when they adopted the imperial gallon, which equals 1.2 U.S. gallons.[46] American whalers continued to use a mix of tuns and gallons for measuring whale oil. A vessel's 'tuns' has since shifted to refer to its volume or capacity for cargo, while the term 'barrels' shifted from being a measure of whale oil to being a measure of petroleum.

When a harpooned whale was alongside the ship, the whale blubber would be cut into letter-sized 'bibles'.[47] These smaller pieces

A cross-section of a whaling ship, detailing the barrel
storage for whale oil and other commodities.

could easily be thrown into the try-pots. Once melted, the hot oil
would be ladled off and poured into a nearby cask, which was
possibly open on top, for cooling. When cool, it would be ladled
or siphoned into other casks, either on deck or already below deck,
for storage.[48] Each cask would be bunged with a wooden or cork
bung.

In the history of European and American ocean exploration, the
whalers followed closely in the wake of seagoing explorers such as
Captain Cook. They found sheltered harbours to resupply their
water and firewood and would start trading with the indigenous
populations for fresh foods. Certain locations around the world
with easy deep-water access and abundant resources became favoured
watering holes. One such harbour was at the town of Russell, in New
Zealand's northeastern Bay of Islands.

Sailing into the Bay of Islands on a Kiwi friend's yacht in the
southern hemisphere's fall of 2011, we moored in Russell's small
bay. There, almost 200 years ago, whaling ships from around the
world would have been moored. Looking towards shore from the
yacht, I could see Victorian-style hotels and houses lining the
waterfront. During the time of whaling, these buildings were the

homes and businesses of whaling provisioners, including some 35 brothels. Now they are carefully painted bright white and elegantly refurbished to cater to hordes of tourists instead of the whalers. According to Russell's tourist brochure, in those early days the town gained the name of 'Hell Hole of the Pacific' due to its reputation as a bawdy and lawless stopover.

Despite the four-plus months it would have taken to sail from America's eastern seaboard, so many American whalers came into Russell during the 1840s to '60s that a full-time U.S. consular agent position and residence was established. The agent would duly record each ship's arrival and departure and assist with other administrative tasks. On our visit, an original historic register was on display in Russell's Museum, opened to the notation for the *Niger*, a New Bedford, Massachusetts-based whaling ship that arrived in 1846. She was already carrying 845 barrels of the highly valued spermaceti oil and 500 barrels of regular whale oil.

When the *Niger* left Russell, the agent recorded that she was carrying only 635 barrels of spermaceti oil. Apparently 210 barrels were transferred to a sister ship, the *California*, which was en route directly to New Bedford. There are two possible reasons for this transfer: whaling captains were known to hedge their bets to make sure that at least some of their prized cargo reached the home port; alternatively, perhaps the *Niger*'s captain sent some barrels ahead with the *California* to take advantage of higher prices, thinking that they might drop before the *Niger* arrived.[49]

Joan Druett has recounted the story of one whaling captain, John Beebe, who, when returning to port with completely full barrels, encountered a pod of whales. Placing the value of the whale oil over temporary hunger, he had all the barrels still containing bread emptied, the bread thrown overboard and the bread casks filled with oil.[50] She related another story of the ship's cooper who also happened to be the ship's captain. In one incident around 1850, Captain John Deblois was setting up the barrels as fast as he could. But his crew was filling them faster than he could set them up. In the chaos of the moving and filling, someone bunged a barrel, in order to move it, with its oil still super hot. The gases in the wood would have expanded, and with no outlet the barrel exploded, drenching and scalding Captain

Deblois with hot oil. Apparently he survived, but was in bad shape for a while.[51]

While other containers and modes of land transportation have certainly been important, wooden barrels and wooden boats have played a critical role in our cultural development and worldwide migration.

# SEVEN

# Organizations: From Guilds to Cooperages

By the thirteenth century, such organizations [guilds and similar]
were well established in cities and towns throughout western
Europe . . . and sought to regulate competition rather than to abolish it.

Edwin S. Hunt and James M. Murray,
*A History of Business in Medieval Europe, 1200–1550*

Guilds were important in, and extensions of,
the tradesmen's social and religious life.

Kenneth Kilby, *The Cooper and His Trade*

## The Rise of Cooperage Guilds

 Unlike the early Mediterranean amphorae workshops, which employed a number of workers, in European history barrel crafting was generally accomplished by a master cooper, assisted by one or two apprentices. Even into the mid- to late Middle Ages, the increasing demand for barrels and larger cooperage vessels from a number of diverse commercial enterprises generated more and more single-cooper workshops, rather than multiple-worker businesses. And paralleling these increasing numbers of individual coopers was the increasing need to organize as trade groups.

The first Celtic and Roman coopers were making but a few barrels per day; apparently this was enough to fulfil the demands of wine merchants and other miscellaneous tradesmen. For these coopers, the earliest incentives to form trade groups or *collegia* occurred in about AD 200 as the Roman Senate began enacting laws to benefit the consumer.[1] However, by the fifth century, these trade organizations had lost their original momentum.[2] They did not seriously re-form until the early Middle Ages, when they developed with the primary goal of promoting religion and/or social purposes. Eventually, more specialized trade organizations, for coopers as well as for many other tradesmen, were established in the eleventh century, with more formal guilds forming by the twelfth century. By that time, the guild's underlining interests had shifted to more commercial concerns, actuated by their desire to control the competition within their local trade.[3]

By the twelfth century, the demand for cooperage had increased substantially, with burgeoning commerce contributing to the reformation of trade organizations. A primary example of the cooperage demand would have been the annual fleet of some 200 merchant vessels sailing from France, loaded largely with barrels of wine for Britain.[4] The records are unclear as to whether a number of individual French coopers, working year-round, could make the numerous tuns and smaller barrels for the thousands of litres of wine, or if this was accomplished by groups of men working together. It does appear that English coopers, at least those making the slack barrels to ship salted fish to France, were formed into companies of several men.

Also by the twelfth century, German coopers were building larger and larger wine-storage casks. Their fabrication would have required a number of men working together, at least for these specific large-cask cooperage projects. The first of a number of tanks of over 200,000-litre capacity were erected within Heidelberg Castle in the fifteenth century, and a similarly large cask was built in Strasbourg in 1472.[5] To accomplish the manufacturing of these large casks and tanks, some loose association of coopers must have formed. Further evidence comes from a print of 1586 that shows four German coopers making large barrels.[6]

There were many interrelated reasons for these organizational developments throughout the Europe that was emerging in this era. While specific goals and concerns certainly varied with the type of trade, as well as within the locale and time period, the main issues were similar for the tradespeople involved: driving their businesses by maintaining relative regional monopolies; providing for their families and themselves in case of disability, old age or death; involvement with special religious and cultural interests within their communities; providing a united front towards governmental demands for time and money, including having a say in the development of rules and regulations which affected the trades; and the need to control the business practices of their members. These goals were

The Heidelberg Tun is a huge wine cask built and rebuilt within
Heidelberg Castle, Germany, shown here in a print of *c.* 1820–70.

not significantly different from those of today's trade organizations;
however, the methods to accomplish them have changed.

## Maintaining Regional Monopolies

To protect their local businesses against encroachment by external
tradesmen, various craftsmen, including coopers, attempted to
maintain near monopolies for their skills and services within their
respective towns and cities. One of their methods was to set the
prices for their products by custom and agreement between the artisans and within the guilds themselves, or through negotiation with
the city aldermen, lords or kings.[7] Today, this would be considered
price fixing, but, for several reasons, it made sense several hundred
years ago. Consider first that everything was made by hand. Product
efficiency could only be achieved by the skill of the craftsman; it
was not possible to purchase a more efficient machine, since there
were no machines! Each product took a minimum amount of time
to fabricate, a time frame which was undoubtedly included in the
pricing calculation. A cooper, for example, could make two to three
average-sized barrels per day. Rushing the production would detract

from the quality, and craftsmen and their guilds attempted to achieve overall excellence.

A second reason had to do with product standards. Product standardization – that is, consistent barrel size for a stated capacity – developed in the Roman era, but it slid a bit until the Middle Ages when the Romans became looser with regards to overseeing and there were ensuing upheavals as new ruling entities jostled for power. However, with cultural and economic well-being slowly improving, consumers started demanding some evidence of the quality of the products and services that they were purchasing. The guilds saw standardization as a way of keeping their customers satisfied, and consistency provided justification for the pricing. To appease their subjects, governments also had a vested interest in ensuring a stand-ardization of barrel containers. Kilby notes that in 1382 an English proclamation required each master cooper to purchase a set of pewter containers sized by 'gallons, potels [a 'pottle' was two quarts], quarts and gills of good and lawful size',[8] which would serve as gauges against which coopers would measure their barrel capacity. These attempts at standardization, along with a cooper's individual trademark – their name or mark – stamped or inscribed upon each item,[9] and the trade organization's setting of the prices, continued through the Middle Ages.

## Religious and Cultural Affiliations

The guilds were not only business oriented, but were well integrated within the social and religious fabric of their communities.[10] Initially, the guilds were heavily involved in religious pageants and festivals and, later, in some charitable activities such as providing for the poor and for the basic education of children. For example, the English coopers' guilds gave money to churches and priests for chantries, an endowment for the singing of Mass and for ceremonies for departed members.[11] By the fifteenth century, the London Coopers' Company had its own altar in the old St Paul's Cathedral. They commissioned a woven cloth pall, resplendently inlaid with gold and silver, which would be draped over the casket of their departed brethren.[12] This close association with the Church is further evidenced by the 1509 motto of the London Coopers' Company guild: *Gaude Maria Virgo* (Rejoice, the Virgin Mary).[13]

Guilds also played significant roles in events for the saints' and holy days. For example, the coopers of Shrewsbury, England, together with other guilds such as the fletchers and bowyers (arrow and bow makers), participated in an annual procession to the local sacred site of Weeping Cross, a march of 3 km or so out of the town, where, as Kilby noted: 'all joined in bewailing their sins and in chanting forth petition for a plentiful harvest'.[14] Apparently in order to assuage a communal guilt over some past mutual transgression committed within their city, the coopers of York, another guild, performed the same play at every pageant: 'Man's Disobedience and Fall from Eden'.[15] But these pageants and festivals did not come cheap; the guilds were expected to contribute money or in kind, which went towards the general and specific expenses of the churches.[16]

As the guilds became established, they constructed and decorated their own meeting halls and raised money for the Church and for their own charities. The London Coopers had, by the seventeenth century,

In this painting of 1738, *The Scullery Maid*, French painter Jean-Siméon Chardin depicts a maid who is using an old barrel, held together with only the upper wooden hoops, as a stand for her washbasin.

established almshouses for some of the poor in their communities and established schools for disadvantaged youngsters.

With no governmental welfare, or accident or disability insurance, guilds provided the only method of raising the necessary funds to provide for the wives and families of coopers in the event of incapacitating accidents or death. Later, as guild funding improved, small pensions were provided when member coopers retired.[17]

## Developing a United Front

With growing populations within the villages and cities, demand for craftsmen's products and services also grew. Recognizing this increasing demand for artisans' goods, the various city and regional rulers and governments saw these as fair game for taxes and for other forms of generating income. To protect themselves against unjust charges, the trade brotherhoods and guilds provided a modicum of negotiating power against the amounts of levies they would be required to pay.[18] In Britain, guilds provided the power to combat, or at least minimize, some of the onerous rules and regulations dreamed up by local governing administrations. For example, although not specifically singled out, the coopers were most certainly affected by directives associated with foodstuffs trades, through which, as Hunt and Murray noted: 'strict rules concerning quality, sanitation, and sometimes pricing were enforced'.[19] As an illustration, in 1267, Henry III introduced the Assize of Bread and Ale, a set of statutes and inspections intended to protect the citizenry by controlling the basic ingredients and production of bread making and brewing.[20] Ale kegs and fermenting vessels came under these regulations. The early English taverns were often situated within homes, with the wife as the brewer. These women, known as ale-wives, would have demanded of their coopers clean barrels and cooperage in the first place and then assistance with keeping the barrels sound and with cleaning used cooperage (by scraping the interior when the barrels went a bit off).

Three centuries later, in 1537, the records indicate that levies, in the form of taxes and materials, were imposed on the various Companies of London (including the Coopers' Company) in order to fit out troops and ships required for the wars waged by Henry VIII.

There were specific charges for the soldiers' swords, daggers, whips, pikes, bows and arrows, horses and saddles, and the coopers were requested to provide barrels for the food supplies. All of this was a significant expense for the some 50–75 coopers then working in London and the surrounding villages.[21] Once more, the collective power of the guilds would have been more effective in dealing with these royal decrees and municipal regulations than individual tradesmen.

Another method by which government entities raised money during the Middle Ages was to charge the trade apprentice a fee once he had completed his four- to seven-year training period. This fee was to matriculate in order to become a Master tradesman and a free-man, a citizen of the particular city in question. France had a similar fee, which was paid to the king, while in England it was paid to the local city administration.[22]

As the guilds came to power, they invariably had to pay fees, bribes and assorted other levies to have a voice within the local political establishment. Councils imposed fees, oaths and ecclesiastical sanctions for disregarding the laws, and sometimes the guilds paid them, and on other occasions they opposed them as best they could. For example, in 1532, the London Coopers' Company paid for 'a pipe of Gascony wine' costing £3 6s. 8d., which they gave to the parliamentary Speaker;[23] in 1536, a hogshead of wine was purchased from a Mr Wood for £1 and was donated to the Lord Mayor of London. Were these unsolicited gifts or bribes? It is impossible to say. In 1541, the Company paid to have a bill drafted so that it could be presented before Parliament; in 1562 it paid to have a statute repealed that set what the members felt were excessively low prices for some of the cooperage vessels made by company members.[24]

## Controlling Business Practices and Members

Not surprisingly, guild organizations had their own internal conflicts. The rising competition inevitably caused some tradespeople to cut corners when making their products. In the case of coopers, some unscrupulous craftsmen made barrels that were smaller than the stipulated capacities in order to save wood; used inadequately seasoned wood to save the cost of the extended air drying; used porous pieces of wood which should normally be discarded; or spent less time on the workmanship, thereby making shoddy barrels which leaked excessively. These defects would be felt most seriously on board ship, with the inherent difficulties of replacement out on the open seas and far away from port. But problematic barrels used in cellars, households and business establishments would have been just as annoying. The formation of guilds indicates that master coopers, like other tradesmen of the day, must have understood the corrosive effects of dishonest or unprincipled coopers crafting substandard barrels. Desirous of protecting their craft and businesses from these and other deleterious influences, they formed brother-hoods, corporate organizations and, later, the more formal guilds.[25]

Within these guilds, strict and sometimes far-reaching rules were developed to ensure the overall quality of cooperage products. These included the prices for the barrels, the kinds of tools which must be used, work hours and the conditions under which appren-tices would work. Master coopers had to pay to get into the guilds, and they were bound by oaths to adhere to the rules.[26] Penalties applied to those who disregarded them. For example, for the coop-ers in France's Champagne district, their Châlon Statutes of 1670 specified that coopers leaving their work for celebrations or other questionable reasons, or eating or drinking during working hours, were to be fined ten sous.[27] And failure to pay dues to the guild was a punishable offence. In 1508, the London Company's officers went to the cooperage shop of a fine defaulter and removed his tools, thereby depriving him of his livelihood.[28] The intent was to embarrass the man in front of his family and friends, who would then be forced to pay his fine. What we might now consider petty activities – fighting among the brethren, being out of working uni-form when in the workshop, working in a stranger's cellar at night and not coming to the election of the local sheriff – were also subject to fines.[29]

Shoddy cooperage also impacted other merchants and guilds. The London Vintners, the guild for those who imported and sold wine, searched out and destroyed all non-standard wine barrels and barrels utilized illegally for other products.[30] While the idea was to shake

The guild seal of London's Worshipful Company of Coopers, founded in 1501. Note the blue adzes and dividers, and the gold hoops, in the centre of the shield.

out unscrupulous wine dealers, coopers who made the sub-standard barrels also faced disciplinary action from their own guilds.

At first, the brotherhoods were local. Eventually, as the organizations became more sophisticated, the state and national governmental agencies became involved. For example, in France the Provost of Paris published, in 1268, *Le Livre des Métiers* (The Book of Trades).[31] Within it was listed the details of France's 121 craft and trade organizations, including coopers and, separately, the stave makers. And in 1444, the French king Charles VII confirmed the statutes for his country's coopers.[32]

Guilds became official in the eleventh and twelfth centuries. In London, the 'coopers' were first noted in the tax records as early as 1296, and by 1501 a charter was granted to the Coopers' Company of London,[33] while the German cooperage guilds were chartered at about the same time. Demonstrating the shift from a religious/social emphasis to more commercial interests, during the Reformation the London coopers' motto changed from *Gaude Maria Virgo* to 'Love as Brethren', with the shield adorned with adzes, dividers and hoops rather than religious icons.[34] In Britain, the guilds were also known as livery companies, and were allowed to wear special uniforms during religious and civil ceremonies. Attendance by the liveried guilds was expected, accompanied by some payment to help defer the ceremonial costs.

## From Cooper to Cooperage

As touched on earlier, despite the centuries of moderate barrel use for wine and other provisions, demand for barrels increased significantly from the fifteenth century, centred largely on the escalating ability of ships to sail to the far reaches of the world. In earlier centuries, while fleets of fishing boats had ventured into the North Sea and off the coast of Europe to catch herring and cod, those voyages were rather short. But as the ships improved, and once they could safely navigate the Atlantic and eventually the Pacific, the floodgates opened, and barrels were required not only for water and food supplies, but also to ship and store the products of those forays – cod, herring, salmon and whale oil. Later, from the mid-nineteenth century, there was the need we touched on earlier for millions of barrels to ship crude oil – the petroleum gushing out of those Pennsylvania wells – to refineries; this is outlined in further detail below. This demand encouraged, and actually necessitated, the further development of cooperages.

Initially, to meet these assorted fish and oil demands, more and more coopers were trained and employed, and they eventually formed cooperages. These coopers, following the actions of their contemporary tradesmen, formed guilds and unions, the efforts of which slowly won wage increases for their members. But by the eighteenth century, with cooperages sometimes employing hundreds of coopers and with the advent of the machine age, management deemed it more cost-effective to install machines, to be operated by less skilled labour, than to hire master coopers to construct each barrel fully. And as the machinery became increasingly sophisticated and prevalent within the cooperages, it slowly replaced more and more coopers.

It is worth here looking in some greater depth at the fishing industry. In the fifteenth century, the Dutch fisheries – known as the 'Dutch gold mine' – moved out of the Baltic into the North Sea, dramatically expanding their herring fishing grounds and subsequently their industry.[35] In the ensuing three centuries, hundreds of coopers were employed in Scottish ports, where the 'Dutch' ships deposited their catches, in order to build the thousands of barrels required to distribute salted fish to consumers throughout Europe. These barrels were slack cooperage made of Swedish spruce and fir with heads of Scots pine, due largely to availability and ease of working the woods. Later, American elm was used for similar reasons.[36]

The successful oceanic journeys of exploration by various nations, and the commercial success of the Dutch, prompted the British to explore other fisheries. They ventured west to Iceland and then on to America in search of cod. And as the early American East-Coast settlers watched the success of the British in their own backyard, they too pushed to develop an Atlantic cod industry. All the while the coopers – European and American – were busy building the thousands of barrels that would package the salted fish, enabling it to be delivered to markets on both sides of the Atlantic. A parallel development occurred with barrels for the whaling industry.

The developing need for barrels for the fishing industry was relatively slow – taking several centuries – compared to the demand that exploded suddenly around the 1860s with the advent of the U.S. petroleum industry. Before this explosion, crude oil usage had basically been limited to medicines, then all of a sudden, there was an insatiable demand for its use as fuel. Kerosene and coal oil came into use for lighting, replacing whale oil, and for heating in cities and homes. Eventually, fuel oil and gasoline became essential as important heating and cooking sources and later as the almost-universal fuel for all modes of transportation in the expanding industrial world. Wooden barrels played a key but extremely short role as containers for both the raw and finished petroleum products. The crude oil was initially transported in barrels from the wells to the refineries via river barges and wagons. In 1859, with only a few wells drilled, the annual demand for these containers was about 2,000 barrels – a number easily met by the surrounding, one-man cooperage operations. However, within just three years, the demand jumped astronomically to over 3 million barrels per year![37] The numerous, but small, operations of the hand-coopers could not satisfy such orders, and the few larger cooperages were struggling to keep up as well. Very quickly, the soon-to-be-large oil producer Standard Oil purchased entire forests and built its own cooperage to ensure a supply of wood and much-needed barrels.[38] Finally, the capacity to meet this demand for barrels to transport the increasingly important liquid gold could only be met by the greater use of newly developed, albeit primitive machinery on cooperage production lines. Cooperage managers could see that machinery had the potential to increase output from two to three barrels per man per day to at least ten barrels per man per day.

This petroleum-barrel story was but a fleeting incident in the two-millennia-plus history of barrels. However, for the American cooperages and their coopers, the impact was both a godsend and a disaster. With the increasing demand, there was work for all. But within ten years, large wooden tanks had mostly replaced barrels as oil containers, and these were then quickly replaced by metal ones – tanks could be positioned on railway cars for transportation to the refineries. Eventually, as drilling rigs became widespread, it was easier to lay down small pipelines to connect the rigs to central depositories, which then shipped the oil via the railways. Wooden

Oil barrels, shown stacked near the wooden storage tanks and oil derricks, c. 1864, in Oil Creek Valley, Pennsylvania.

barrels held on for a few more years, taking refined oil and gasoline to end-users. However, because of the viscosity of these refined products, barrels needed to be made from 'clear, sap-free, white oak', coated on the inside with water-soluble glue.[39] Unfortunately, due to inconsistencies when applying the interior coating, and a lack of quality control to fix problems, the resulting leakage was unacceptable to gasoline and oil dealers. These flaws and the resulting seepage became a significant factor in pushing the refineries to make the switch from wooden barrels to metal drums.

By the late eighteenth century, substantial improvements in iron casting, steel manufacturing and precision machining led to specialized machines being built to fabricate a huge range of industrial, consumer and agricultural products. With wooden cooperage still an important container for a myriad of products, machines were also built for almost every aspect of barrel production. Despite production-line coopers rallying against anything which was eroding their livelihood, the competition to produce increasing quantities of inexpensive barrels, whether for oil or other commodities, drove cooperage owners and managers to install machines wherever possible.

The need for petroleum barrels was phased out relatively quickly; however, great demand still existed for containers for other products, such as whiskey, beer and sugar. The late eighteenth and early nineteenth centuries were the heyday of cooperages. Hundreds were functioning in North America and Europe, and a number employed upwards of several hundred workers. But the experienced coopers' days were numbered; steam power, and later electricity, became readily available and replacement by machines took its toll.

The earliest machines to impact barrel making were ones that planed the rough staves and heading, cut the circumference of the heads and 'trussed up' the barrel (brought the staves together) with a steel cable. Later, a 'buffalo' for driving the hoops (named after the New York State city where they were made), a machine to cut the ends of the barrels and a jointing machine were refined and added to the production lines. As skilled coopers retired, production-line employees took on the jobs of feeding the machines, rather than building complete barrels.

Simultaneously, power-driven saws were also introduced in stave mills, including one in America called a cylinder or barrel saw. This was a cylinder with teeth on one end which cut rounded staves from the quarters of the stave bolts. The barrel saw was used up until the late 1900s and has now been replaced by band saws.

All the while, cooperages on both sides of the Atlantic were busy making more and more barrels for whiskey and beer. Britain had several large cooperages producing barrels primarily for beer, as well as other commodities. One, the Bass Brewery at Burton-on-Trent, was by 1889 employing about 400 coopers.[40] With a production line powered by steam, records indicate that they were manufacturing about 2,000 barrels per week. By contrast, today's two largest American bourbon-barrel cooperages each employ half that number of coopers and produce 2,000 barrels per day. Although, to be fair, making beer barrels involved a few extra steps – such as pitching (lining) the interior of the barrel to cope with the additional pressure and installing a brass bung to tap the beer. Beer barrels are typically smaller, and the staves and heading are much thicker, to contain the pressure and rough use.[41] Plus cooperages today have the advantage of far more sophisticated machinery to fabricate almost every step of a barrel, as well as conveyors of various types to move the barrels automatically from one processing station to the next.

In the U.S., with whiskey production averaging over 80,000,000 gallons (3,028,400 hectolitres) per year between 1880 and 1918, there was an increasing demand for containers to age the distilled corn or rye mash. Prohibition, which lasted from 1920 to 1933, halted most whiskey production and in turn put a serious dent in the cooperage industry. But by 1936 whiskey production had recovered and slowly climbed to over 220,000,000 gallons (8,327,910 hectolitres), requiring a peak production of over 4.6 million barrels annually.[42] About twenty large cooperages, located primarily in the Midwest, close to the sources of white oak, and numerous smaller cooperages established in almost every city, were involved in these efforts. Despite the hundreds of men employed, they were still a small contingent within America's workforce.

Barrels for sugar were another important product for some cooperages. In 1883, the California Barrel Company was established within the developing industrial area of Potrero Point in San Francisco, largely to fulfil the demands of the adjacent Spreckels Sugar Refinery.[43] For those slack barrels, they sourced elm and basswood from the eastern United States. Typical of many cooperages, California Barrel also made barrels for beer and wine, sourcing oak from the southeastern states, and made woodenware – pails, tubs and slack barrels – from spruce, Douglas fir and hemlock sourced from northern Californian forests and cut into staves and heading in their huge mill in Arcata, California.[44] California Barrel was phased out in the early 1900s as paper sacks for sugar packing replaced the barrels, and their woodenware lines were supplanted by other materials.[45]

Over the several hundred years during which cooperages were created and expanded, coopers moved from being members of craft guilds to cardholders within trade unions – all the while attempting to hold on to jobs and maintain pay levels. Mimicking the history of other labour unions, some cooperage strikes turned violent and accomplished little. Slowly, as the demand for barrels decreased, the coopers realized that they were waging a losing battle. Today, in the U.S., just five large bourbon-barrel cooperages, fewer than ten wine-barrel cooperages and some small number of related cooperages continue to exist. Great Britain has a few cooperages processing barrels for Scotch and whiskey, and even fewer single-cooper operations. France has about 50 wine-barrel cooperages, most currently operating relatively successfully due to the increased worldwide demand for wine barrels over the past 30 years. Spain has several wine and a few sherry cooperages and Portugal some cooperages for port. Italy has a couple, as does Germany, with the cooperages in both countries primarily fabricating the larger casks. There are fewer than 50 throughout all of Eastern Europe and Russia. Australia has five for the wine industry, while South Africa and Chile both have a few and China has at least one. Japan and Taiwan have a few for sake and whiskey barrels.

## American Cooperage Associations

Early American coopers in the seventeenth and eighteenth centuries were, much like their European counterparts, a comparatively

These French workers are loading wine barrels onto a wagon. In the days
before forklifts, a ramp was employed to help raise the heavy barrels.

independent lot. However, with relatively few operating within each
town, their ability to form guild-type organizations was limited.
These single-man operations were relatively short-lived, with the
multi-men cooperages developing somewhat quickly in the mid-
1800s. The coopers employed within those companies attempted to
form unions, while the owners joined trade associations.

One of the early trade associations was the Tight Stave Manufac-
turers Association, created in 1897.[46] The present-day Associated
Cooperage Industries of America is an outgrowth of that association.[47]
Formed in 1934 and still operating, this organization now encom-
passes stave mills, cooperages (primarily the ones for bourbon with
a few wine-barrel members), cooperage machinery producers, U.S.
bourbon distillers and Scotch whisky distillers and coopers, as they
are the major purchasers of used bourbon barrels.

For working coopers, especially those employed in larger operations during the late nineteenth and early twentieth centuries,
pressure from the cooperage owners to lower costs through the
increasing use of machinery, and the subsequent wage erosion, did
not endear them to each other. Attempts by workers to organize as

unions met with limited success, as their numbers were not great when compared to other industries. For the same reason, they did not benefit greatly by becoming sub-unions of the more powerful national unions, such as the A.F.L.-C.I.O.

I managed a cooperage in Kentucky which had a semi-autonomous and unofficial union. Still smarting from some previous working problems, the employees seemed hostile to almost any change. Only by sitting down with them and working though all the issues – wages, benefits, productivity goals and working conditions, including training for the newer, more sophisticated machines required for making modern wine barrels – did we achieve a mutual understanding and acceptance. This strategy was fruitful for both the company and its employees: the company experienced significant improvements in operational efficiency and subsequent profits, the quality of the wine barrels improved to the point where they were in demand globally and the workers' wages and benefits were increased.

As with many businesses, one-person cooperage operations have either ceased to exist or have merged to become larger – a process begun with the increased demand for cooperage (especially for crude oil) in the eighteenth and nineteenth centuries. Some small operations have managed to survive, however, and on a recent trip to the Gers, in the Armagnac region of southern France, I visited one fellow who was doing quite well – his workshop was spacious and his lovely house nearby had a swimming pool. He was automated enough to produce three Armagnac barrels per day and seemed to have a steady client base.

At the other end of the spectrum are the giant cooperages for bourbon barrels. The two largest operating today are the Blue Grass Cooperage in Louisville, Kentucky, owned by the Brown-Forman company of Jack Daniels fame, and the two privately owned World Cooperage plants – one each in the towns of Lebanon, Kentucky, and Lebanon, Missouri (you can bet that causes confusion with deliveries!). Individually, these plants produce upward of 2,000 or more bourbon barrels per day. This is on a par with early large U.S. and European cooperages making the millions of barrels for oil, sugar, molasses, beer, flour, fish and hundreds of other products. As can be imagined, these cooperages were at the forefront of mechanization, although, with unions agitating for lost jobs, not necessarily advancing as speedily as the owners desired. Today's wine-barrel cooperages are somewhere in between these extremes – ranging from 20 to 300 barrels per day.

## The Beginning of the End

In the United States, the first significant indications of the wooden barrel's demise started in about 1905, when flour producers found it cheaper to ship the flour in cotton or paper sacks without placing the sacks in barrels – previously a common method.[48] With the less expensive sacks, and the barrels being more cumbersome, millers and their patrons slowly switched within the course of just a few years, leaving the flour-barrel coopers looking for other customers.

As late as the 1980s, some wine was still being bulk-shipped in barrels to Bercy, a wine entrepôt in eastern Paris with its wharfs along the Seine. But, as I observed on visits during the 1980s and '90s, this well-established corner of the wine trade was also in transition. No longer were the barrel-delivery barges tied up at the Seine wharf; wine in barrels was delivered by rail and truck or the wine was delivered directly in stainless-steel tanks via trucks. Instead of delivering barrels to the retail outlets, trailer loads of empty glass bottles were being shipped into Bercy, filled in the warehouses and then loaded on to trucks for delivery to local restaurants and wine shops.[49] Eventually, even the bottling operations would cease, shifted to the wineries or sites closer to the vineyards. With all this change evident even in the 1980s, locals were planning for a restoration of the area. By the 1990s, most of the old wine warehouses had been restored as shops and restaurants, leaving only the railway rails embedded in the cobblestone streets and establishments with names such as Restaurant Chai 33 and Claret as remembrances of the golden years of the wine trade.

Beer, a staple in Europe and America, was also initially delivered and dispensed from wooden barrels.[50] Towards the end of the twentieth century, however, the beer industry yielded to the improvements provided by steel, then aluminium, kegs and eventually cans and glass bottles.

The de facto end of the wooden barrel as a bulk shipping container occurred on 26 April 1956, when a converted tanker ship, the *Ideal-x*, sailed out of New York harbour bound for Houston, Texas. On board that ship were 57 'containers' – forerunners of the modern 6- and 12-metre-long metal boxes now carried on trucks, ships and railcars as they move between industrial facilities the world over. This shipment of containers wasn't exactly a first: truck trailers had been carried on railcars and ships, railcars had been transported on oceangoing ships and barges and metal box containers were in use. What was unique, however, was that these were the first containers specifically designed to be shipped interchangeably by boat, truck and rail.[51] This was the start of a shipping revolution called intermodal shipping.

The concept and implementation of intermodal shipping containers was the brainchild of Malcom P. McLean, a trucker who, frustrated by the delays in dropping off and picking up his cargo in the New Jersey ports, dreamed of a more efficient way. Most of his freight was break-bulk – that is, small, individual items of cargo that required much handling but which could also be consolidated into larger loads. His goal was that items would no longer have to be individually handled numerous times during transit; only the container itself, loaded with numerous goods, would need to be shifted between transportation modes. And no longer would the truck drivers have extended waits to pick up their cargo: a crane would unload or load a container off or onto the truck in a few moments, and the driver and rig would be on their way. In a busy port, thousands of containers could be shifted every day by a few people working cranes and giant forklifts, vastly increasing the amount of cargo which moved in or out of a port and reducing the transportation costs.

McLean had actually started his trucking career hauling empty tobacco barrels. So he knew that loading or unloading numbers of barrels from trucks, even with skilled longshoremen or tobacco-factory workers, took an hour or two once the truck was backed up to the loading dock. He was also frustrated by the other, often longer, delay – just waiting to get his truck actually up to the dock! Similar delays were experienced when moving freight and cargo between trucks, ships and railcars.

McLean's concept put a number of intermediary workers – the longshoremen who physically moved the freight from one form of transportation to another – out of work. But it vastly improved freight handling, lowered freight costs, improved security for the cargo and exploded global trade. Containerization basically put one of the last nails in the coffin of the use of wooden barrels as multipurpose containers.

# EIGHT

# Oak: Wood for Barrels

The oak genus, *Quercus* . . . constitutes over 400 species,
of which a few are widely used to construct barrels for wine.

Geoffrey Schahinger and Bryce Rankine, *Cooperage for Winemakers:
A Manual on the Construction, Maintenance and Use of Oak Barrels*

 Wooden barrels, despite their many similarities to wooden boats, have undergone far fewer changes over the several millennia of their respective evolutions. Today's barrels appear much the same as those images captured in ancient friezes and mosaics from thousands of years ago. A Roman dockworker or Celtic housewife would easily recognize today's wine or whiskey barrel. And if a cooper from 2,000 years ago could step into a time machine and be warped into a modern cooperage, while he would no doubt be perplexed by the machinery, he would be able to understand each step of the barrel-crafting process. By comparing the machine processes to those of his own hand labour, he would rapidly see that little has changed. However, with the advantage of hindsight, we can now observe some of the subtler details which have evolved with wooden barrels: how some of the New World (North American) oaks are incorporated into specific barrels; the dramatic shift from hand tools to automated machinery in the interests of mass production; the change from organic hoops to steel hooping; the greater attention paid to the oak-fire toasting of wine barrels, which enhances the flavours in oak-aged wines; and the explosion of commodities which have been shipped, stored and otherwise contained in barrels throughout the centuries.

Without being too technical, let us examine first the raw materials – the timber. Over the ages, many woods have been tried and utilized; however, oak was and is the primary timber for tight barrels, including wine and whiskey barrels.

## The Forests and the Oak

For the current wine market, wine barrels are made from woods of the white oak family, which are sourced in Europe and North America. The main European white oak species is *Quercus robur* (English oak), under which are two primary subspecies: *Q. sessilis* and *Q. pendunculata*. These subspecies have a wide and overlapping range throughout Europe. In North America, the main oak for cooperage is *Q. alba* (white oak), also with several subspecies and coinciding ranges.

The price of a European oak wine barrel – in the u.s., for example – is several hundred dollars greater than one made of American oak. This factor alone indicates that there are some significant differences between the oaks. The two most important are the flavours imparted by the oaks to wine and the methods required to produce the staves – the first more of a personal preference and the second a mechanical trait. European white oaks are known for their subtle nuances of vanilla and spice flavours, which they can lend to wines with which they come into contact. On the other hand, American white oak has more pronounced flavours of coconut and vanilla, typically associated with the tastes of bourbon and whiskey. While these flavours are often strong in American oak, they can be toned down by an extra year of air-drying time. For the world's winemakers, having this choice of flavours, in a range of nuances, allows them to fine-tune their wines, creating their own unique styles.

From the cooper's point of view, the other significant difference between the oaks of Europe and North America – *Q. robur* and *Q. alba* – is how the staves are extracted from the log bolts. American white oak has an inner cellular structure known as the tyloses, which block the flow of liquid. This structure allows the staves to be sawn directly from the log bolts, improving stave productivity and yielding a greater percentage of usable wood. By comparison, to produce staves for wine barrels from European oak, the log bolts should be split, mechanically or by hand, as one would when splitting firewood. Once split, the roughly triangular piece must then be trimmed to a rectangular shape in order to form the stave. These splitting and trimming processes are rather laborious and less efficient in terms of yield, thus adding to the cost.

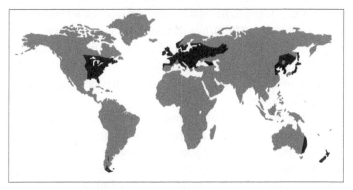

The world's major deciduous forests (dark grey) occur in
roughly the same latitudes in both hemispheres. Oak trees
are not indigenous to the southern hemisphere.

Be they the American or European oak varieties, the majority of
trees used to make wine and bourbon barrels grow in the northern
hemisphere's mid-latitudes across the middle and eastern United
States and Western and Eastern Europe. As the French cooperages
have done an incredible job of branding their wine barrels and their
oaks to ensure that their cooperage is highly sought after, we will
discuss those barrels first.

## European Oak

The two main European white oak subspecies, *Q. sessilis* and *Q.
pendunculata*, are prevalent throughout the European deciduous
forests, from England stretching east into Russia. Under the direction of the French government's Office National des Forêts, the
extensive oak forests within France are generally agreed to best
meet the wine cooper's needs. The French forests, especially those
in the colder northerly regions, are well suited to growing the trees
slowly, and it often takes up to 200 years for them to gain sufficient
girth for harvesting. This slow growth results in timber with a
closely spaced wood grain, and it is these tight-grained trees which
have more extractives than the wider-grained wood of the southern forests. When used in a wine barrel, the wine interacts with
this tight-grained wood in a slow, controlled manner, making it

ideal for prolonged barrel maturation and superior integration of the wine flavours. The Office National des Forêts nurtures the trees throughout their long life by keeping the forests dense, so that the trees grow straight and tall, and by thinning the lower branches to minimize knots – producing extremely high-quality and expensive logs.

Within the French and European forests, the two main oak subspecies are intermixed and some hybridization has occurred. Usually, however, one subspecies or the other will predominate within any given forest. Each subspecies tends to have its own unique 'flavours', amplified by their terroir, or regional growing conditions. Winemakers will often request wood from one forest or another to obtain more or less of the nuanced flavours desired for their wine style. The expression of a terroir in oak wood, like that of a vine-yard, will come through in the way in which the wood's flavours are influenced by the ecology of the forest site: the climate; the soil and its minerals; the altitude; and the compass orientation of the land, such as a north-facing slope. Also, wood-grain tightness is now one of the key factors which winemakers consider when purchasing barrels – so much so that some believe that grain tightness is more important than the actual source-forest location.

In France, once the forest management deems the trees ready for harvest, they allow all interested buyers – coopers, furniture makers, cabinetmakers and veneer producers – to inspect the standing trees. They do, in a sense, operate blind: able to see only the bark and a few exterior knots, it is difficult to identify potential interior problems. They must rely on their years of experience. The true measure of the tree's worth is exposed only once it is felled. The inspector's second examination, of the cut ends, can reveal any rot in the centre of the tree or cracks created during serious wind storms or from the felling process.

Once felled and trimmed, the logs are then sold at auction, and those purchased by coopers are transported to their mills to be split into the staves and heading for drying. While on a recent boat trip on the Saône River near Dijon, France, I encountered a barge loaded with logs for a cooperage. In my limited French, I learned from the couple in charge of the barge that the logs had been harvested in France's

The extent of European white oak (*Quercus robur*), indicated by the dark shading.

Vosges forest region and were headed downstream, destined for a cooperage near Beaune that had purchased them at an auction.

French coopers and winemakers have generally preferred oak from their own forests. A twentieth-century French cooper, Jean Taransaud, wrote: 'Les pays producteurs sont notamment . . . l'Amérique dont les bois on l'inconvénient d'être dépourvus de tanin' ([Of the white oak] producing countries, notably . . . American wood has the disadvantage of being devoid of tannin).[1] Transaud felt that American oak has too little tannin to hold up to the robust and refined French wines. Others believe that American oak has too much tannin. Nevertheless, the tannin of American white oak can be ameliorated through longer air-drying of the wood. Also, in difficult economic times, some French cooperages have purchased American oak staves, with documented shipments as early as 1900 for a load of oak staves shipped from Florida to Cette, France.[2] Recently, in order to try to remain economically competitive, some French wineries have purchased American oak barrels – which are roughly half the price of those made from French oak.

## American Oak

American mixed temperate forests, where the oaks grow, are huge, covering approximately 53.8 million ha in the central and eastern United States,[3] although not quite as large as the total of Europe's at 70.4 million ha. Some 90–122 million board metres of white oak are harvested each year – a minor portion of the total 51 million cubic metres of the total annual hardwood harvest.[4] And from this total amount of oak harvested, the oak selected annually for cooperage use – both for whiskey and wine barrels – is an even smaller 10 per cent.

While some wood is harvested from government-owned forests, the majority, about 60 per cent, comes from the private wood lots which dot the eastern United States. Here the white oak is just one of a number of intermingled hardwood species. The forest owner hires a logger, who will usually harvest all the mature trees at any one time. Once felled, the logger/timber harvester will take each individual species to the specific mill which processes that kind of lumber – that is, hickory for tool handles and cabinetry, walnut for cabinetry and gun stocks, red oak for flooring and the white oak for both wine and whiskey barrels. Besides barrels, the oak's uses include furniture, kitchen cabinets, plywood veneer for dining- and

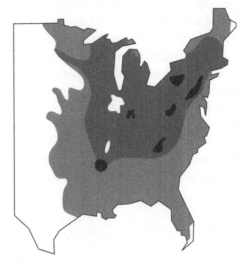

The extent, and density, of American white oak (*Quercus alba*), the major timber species utilized for whiskey barrels, and for some wine barrels.

living-room cabinets, flooring and, surprisingly, pallets. Hardwood pallets are in demand for heavy equipment and for pallets that are reused – so much so that in the 1990s, 50 per cent of the white oak harvested went to make pallets. Out of the more than 40 species of oak in eastern U.S. forests, only white oak, *Q. alba*, and a few of its subspecies are used for wine and whiskey barrels.[5]

The demand for *Q. alba* oak for whiskey barrels far exceeds that for wine barrels – typically 800,000–1,000,000 whiskey barrels are made each year versus some 200,000–300,000 wine barrels. And while winemakers do request wood from certain states, there are far too many microclimates to offer any reasonable specificity. Currently, the stave-mill site and grain tightness are the primary selection factors for winemakers requesting American-oak barrels.

Typically, the oak trees utilized for American-oak wine barrels are not suitable for harvesting until they too, like their French counterparts, are at least 100–200 years old. This is for practical reasons: cut the trees too young and they would be of insufficient girth to make them economically viable to process; wait too long and the trees will start to suffer the defects created by years of buffeting winds and storms. Battering by strong winds creates cracks in the wood: crevices which allow insects and microorganisms to gain a foothold within the tree's internal passageways, creating rot and other defects and reducing the usefulness of the wood. As the trees with the tightest growth rings are the most desirable, the slow growth pushes the wood for barrels towards the upper limits of the age range.

Girdle a tree and it dies. This is because a tree's growth rings are formed by the cambium-layer cells – the active, growing cells that are just underneath the bark. These cells are the vessels that transport water and nutrients from the roots upwards to the leaves and return energy molecules, created within the leaves by photosynthesis, back down to the root zone. With the increasing warmth and moisture in the spring, these cambium cells grow rapidly, creating what is called the early wood. During the summer, the growth slows down and is termed late wood. Despite being made from tiny cellular structures, early wood and late wood create alternating light and dark bands that can be seen with the naked eye and appear as the tree's growth rings.

Each year, the cambium layer produces an exterior set of cells just inside the bark. At the same time, its interior cell layer becomes the heartwood, or dead but structurally sound wood. As this transformation occurs, the cells in Q. *alba* develop tyloses to effectively close off and block the flow of liquids. The new, outer cambium cells take on the transportation role. It is this heartwood which is used for the staves and heading, and this process explains one of the reasons why the cambium-layer cells (also called sapwood), without the yet-to-be-developed tyloses, are removed from staves and heading destined for wine barrels.

Like any plant, if a tree is stressed – growing on a thin-soiled hillside, particularly a north-facing one (in the northern hemisphere), or at a higher elevation or latitude with a shorter growing season – it will grow slowly, resulting in tight, 1-mm or less growth rings. In the United States, oaks growing in northern Iowa, northern Ohio and southern Minnesota are typically subjected to these conditions. The Tronçais region, in the centre of France – the region which was protected by Louis XIV's Secretary of the Navy – has slow-growing trees because of the altitude (300–400 m) and latitude (47°N). Many other regions in both America and Europe also grow highly suitable oaks for wine barrels.

For the American loggers harvesting the trees, those most desired, at least from a yield standpoint, are those from dense forests. It is within those woods that the trees are forced to grow vertically – straight and tall, with few lower limbs. With little sunlight reaching the lower canopy, their leaves reach upwards towards the illumination rather than branching out laterally, which they would do as a single tree in a field or on the edge of a forest. Unfortunately, these forests are harder and harder to find as logging roads, power-line right-of-ways, pastures and paddocks continue to make incursions into the pristine woodland.

While sapwood and knots might be acceptable defects in wooden barrels destined for other products, they are not suitable in wine barrels. With their potential to cause leaks and create off-flavours, they must be rejected in the staves and heading. Other obvious defects often result from human incursions into the forests: nails; fence wire and the staples and nails used to hold it in place; horseshoes hung on nails or branches; rocks left in the crotch of a limb; and electrical insulators embedded in the wood. All of these create havoc if they are still in the logs when they are cut in the sawmills. Therefore, oak trees growing along fence lines or around old cabin sites are particularly avoided. Unfortunately, the saw miller often sees these items *after* the saw cuts through them, resulting in a damaged or ruined saw blade. Along the same lines, certain European forests are avoided as they have far too many bullets and projectiles embedded in the wood, remnants from the First and Second World Wars.

Other defects that the coopers try to avoid are mineral streaks, which occur as the tree takes up nutrients and can, over time, be a conduit for leakage; shakes, splits and cracks, which – as already mentioned – occasionally occur when the tree is felled or has borne the brunt of extreme weather; twists, if the tree grows irregularly; and rot, which results from insects and bacteria attacking a defect in the tree.

In years past, once the desired tree was chosen, the woodsmen would fell the tree with an axe or large-toothed cross-cut saw, cutting just a few per day. Nowadays, utilizing a chainsaw, one man can fell many multiples of those numbers. After felling, the trees are cut into long sections, typically between 3 and 6 metres. This length is determined by what is convenient to haul to the mill and by a multiple of the desired length of the future staves. For example, if the staves are to be 1 metre long, after the woodsman fells the tree, he would cut a 3.5-metre length. The stave mill would then be able to obtain three 1-metre stave bolts, with a small section of extra length to allow for some waste and shrinkage or to become pieces for the barrel heading.

The height of many trees allows for a second cut, which may, or may not, be used by the cooper depending upon the quality (usually determined by the number of knots or other defects). If rejected by the cooper, it would be sold for making furniture, flooring or pallets, or to other woodworkers who can work around the knots.

## Staves: Quarter-sawn versus Split

In the era prior to trucks and mechanical loading equipment, the log bolts would be processed right in the forest. There, they were cut to stave length with a cross-cut saw, then split and trimmed into staves to be hand-loaded onto wagons or river barges for transport to the nascent cooperages. With today's powerful skidders, loaders and transportation equipment, whole trunks are easily transported to centralized mills. Once at the stave mill, the trunks are cut into the respective bolt lengths. For the American white oak, the bolts are then sawn to extract the staves, while for the European white oak – as previously mentioned – they are split. With either sawing or splitting, the bolts are first halved lengthwise and then quartered.

Quartering is a relatively unique method of wood processing. It shifts the wood's grain orientation and thereby facilitates bending the staves. With the proper amount of heat applied, a quarter-sawn or quarter-split stave can be curved and will not break or crack.[6] The limited range of wooden items utilizing quartered timber includes certain types of flooring, special sections of furniture, musical instruments and barrel staves. For comparison, dimensional timber, as in pine and fir for the studs, joists and beams in a house or the redwood paling for a wooden fence, are flat-sawn. Flat sawing consists of a series of parallel cuts right through the log. It is done for speed and economy, regardless of knots or grain patterns.

Quartering has the additional advantage of producing wood with less cupping – curvature across the width of the board – and shrinkage. The end result is that quarter-sawn or split staves are relatively stable and less likely to warp or move once assembled in the barrel – qualities which are essential in order to maintain the alignment of the staves snug against one another, thereby preventing leakage.

For the cooper, looking at the end of the log bolt, the quarter sawing or splitting process takes place along the medullary rays, the cells which branch out laterally from the centre of the tree. These rays transport the water and nutrients to and from the cambium layer to the branches. Rather than a grain pattern, what you see when looking at the flat surface of a stave are the silvery sides of the medullary ray's

Quartering oak logs;
this is the step prior to
extracting the staves
for barrels.

vascular cells, which run up and down a tree. Ideally, the cooper wants the annual rings of the grain visible only at the ends of the staves; otherwise there is the potential for leakage. A winemaker can also check for grain tightness by observing the stave ends.

As we know, early Celtic coopers split the log bolts using a heavy maul and wedges, or froes. This would have been easier than hand-sawing each stave out of the bolt. The bolt is first split in half lengthwise, and then those halves are split again, into quarter rounds – thus the term 'quarter'. Looking at the top of the quartered log bolt, the staves are then split out as elongated triangles. It is from these triangular sections that the staves were shaped into long rectangles with draw knives; nowadays band saws are used to trim away the waste, making the task somewhat quicker and easier.

In America, the process of quarter sawing evolved by trial and error once the European coopers arrived. They knew, from their experience with European timber, that sawn staves could leak, as the sawing process does not always follow the grain of the wood. So, by mistake, or trial and error, sawn staves were placed into barrels. However, a few mistakes like this over time, and an observant cooper realized that, if the wood was American white oak, and then from only a few particular subspecies of white oak, it generally did not leak. However, he may not have known why. We now understand that only certain species of American white oak have the tyloses. And it is the tyloses' ability essentially to shut down the transportation capabilities of cambium-layer cells that allows the use of staves sawn, as opposed to split, from those select species.

The major advantage of sawing versus splitting is yield: the sawing process typically generates approximately 50 per cent of usable wood, versus roughly only 30 per cent by the splitting process. And with the demand for more barrels, and pressure to make staves efficiently and inexpensively, American coopers sought ways to cut the staves rather than split them. The sawing process follows roughly the same game plan as for splitting – up to the quarter rounds. At this point the staves are sawn, alternating off one of the flat sides of the quarter round and then the other until the staves become too small for use.

The realization that American white oak could be sawn was an incentive to mechanize the process. In 1870 the cylinder or drum saw was developed.[7] It cut latitudinal curved staves out of the quarters, saving on having to do the extra steps of hollowing the inside of the stave and rounding the outer surface. The drum saw was used extensively up until the 1990s and was then replaced with band saws for cutting straight-sided staves from the quarters. With the use of super-tough, ultra-thin band-saw blades, the yield is actually better than with the drum saw. For coopers on the eastern side of the

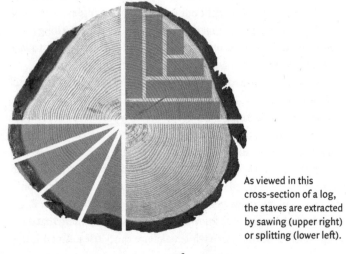

As viewed in this cross-section of a log, the staves are extracted by sawing (upper right) or splitting (lower left).

Atlantic processing European oak, the jump to mechanization came in the form of hydraulic splitters. Most European coopers now use huge, mechanical wedges to ease their labours for splitting the staves, often while engaging computers to help maximize the yield, and then band saws to trim the staves.

Since the advent of sawing staves for liquid-tight barrels, coopers and winemakers have argued over its merits and drawbacks. The significant yield from sawing staves out of a log bolt versus splitting them is not in dispute. But the sawing process goes in a straight line and does not always follow the natural flow of the wood's grain. A sawn stave may be cut 'across the grain', so to speak, possibly exposing some of the cell interiors to the wine in the barrel. Some winemakers believe this allows additional exposure to the tannins and chemicals within the wood and is a negative attribute of ageing wine in American oak barrels.

These taste differences seem to rely more upon individual perceptions than scientific fact. Certainly the barrels made of split French oak age some of the most desirable wines in the world. However, there are also a number of wineries around the world – Ridge and Silver Oak in California, Spanish wineries such as Bodegas Muga and Vega Sicilia, Australia's Penfold Grange and Chile's Cousino-Macul, to name but a few – which have achieved both financial success and huge cult followings by ageing their wines in barrels made with American oak.

We have not yet talked here about the boards required for the heads of the barrels, as obtaining the longest pieces of clear, defect-free wood – that is, for the staves – is the most difficult task, and that is where the focus is. With the fewer pieces of heading required for each barrel, they appear, almost magically, out of the sections of staves from which a defect needs to be cut, or from tree trunks which have an extra bit of length at the end after cutting the bolts for the staves. These smaller pieces are processed like the staves in that they are either split or sawn out of the bolts. Also like the staves, they are then stacked for drying.

## Air-drying the Staves and Heading

In recent years, many summertime family events have become focused around the outdoor barbecue placed on a wooden deck, and decking in general has become very popular, leading to a huge demand for (ultimately) dry wood. The deck installer knows that when using unseasoned wood he can place the decking very close together; it will then dry and shrink to make an acceptable gap between the boards. Just the opposite is necessary for barrels, furniture and joinery. The wood needs to be dry prior to being placed into a barrel to avoid further shrinkage, but it still needs some moisture in order to bend without cracking. Typically for wine barrels, European oak is dried to 14 per cent moisture content, while American oak is dried to 12 per cent.

This 12–14 per cent moisture level is a balance between providing enough moisture left in the wood to allow it to be worked without splintering, yet not enough to shrink further when the barrel is completed. In places like France and the eastern United States, with their higher humidity, these ideal levels can't always be achieved in two years. Longer air-drying, or the use of artificial drying in dry kilns, is sometimes necessary. French cooperages learned this lesson in the 1960s and '70s, when they were just starting to build and deliver significant quantities of wine barrels to the drier climates within the budding Californian and Australian winemaking markets. The 14 per cent moisture level was fine given the humidity of the French countryside, but for barrels exported to much more arid places, left in their dry-climate warehouses or out in the open before being taken into the winery cellars, conditions would often shrink the staves further and loosen the hoops. It was common in those days to recommend two to three days of soaking with water prior to using for wine to compensate for the expected shrinkage. Now, with the barrels wrapped in plastic to conserve the moisture and stored in climate-controlled warehouses, shrinkage is rarely an issue.

To dry naturally, the staves and heading need at least two years in the open air for the sun and wind to achieve the moisture levels required for liquid-tight barrels. During the drying period, the moisture is evaporated from both within and between the wood's cells. The

A typical stack of staves drying in a cooperage yard. The staves are angled to allow the rain and snow to drain.

sun and wind speed up the process, while rain and snow slow it down, but do help by washing the wood and thereby leaching out tannins. By reducing the tannins, the compounds which cause harshness and astringency, the wood flavours are softened – which is especially desirable when making wine barrels. With the more pronounced tannins in American oak, an additional year of air-drying is beneficial to further decrease the tannins.

Once split or sawn, the staves are stacked in ricks to promote airflow around as much of the stave surface as possible. Stave-making regions each have their own unique stacking formats and styles. They all have, however, the same basic requirements: to minimize the contact surfaces of each piece of wood against one another in order to avoid mould developing in the overlaps; to slant the wood to allow the rain to run off; and to orient them for maximum exposure to the flow of the prevailing wind – or in the case of some of the older European methods, such as stacking in the form of a beehive, to create their own upward draught.

# NINE

# Air, Water and Fire:
# Crafting Wooden Barrels

As other people have a sign,
I say – just stop and look at mine!
Here, Wratten, cooper, lives and makes
Ox bows, trug-baskets, hay-rakes.
Sells shovels, both for flour and corn,
And shauls, and makes a good box-churn.
Ladles, dishes, spoons and skimmers,
Trenchers too, for use at dinners.
I make and mend both tub and cask,
And hoop 'em strong, to make them last.
Here's butter prints, and butter scales,
And butter boards, and milking pails.
N'on this my friends may safely rest –
In serving them I'll do my best;
Then all that buy, I'll use them well
Because I make my goods to sell.

Sign in front of Wratten's Cooperage Shop
in Hailsham, Sussex, early twentieth century.
From Kenneth Kilby, *The Cooper and His Trade*

 Regardless of the type of oak, making quality wine barrels is a process of applying air, water and fire to the wood – all under the cooper's watchful eyes.

After several years of sitting quietly in the open air, the rough and weathered staves are now sufficiently dry. They are taken into the cooperage, where the first step is to plane them to a consistent thickness, a normal process in most woodworking activities. Particular to a cooperage, this planing also rounds the outside of the stave to enhance the exterior curvature of the barrel and cups the inside to aid in bending.

A stave from the drying yard is not sun-bleached white, as is wood often found on a beach, but has been darkened by mould and airborne dirt particles. With its rough exterior surface from splitting or sawing procedures, combined with the darker colour created during ageing, any defects on the staves are difficult to spot. Therefore, removing the sun-and-rain-darkened surface wood, and smoothing the wood by planing, provides an additional benefit: the cleaned wood facilitates further inspection. The cooper will then be able to check for damage missed at the mill and possible damage incurred during the drying process.

With each stave now sporting a concave inner surface, a convex outer surface and a clean bill of health, they go to the jointer. In the jointing process, an angle is cut on each side edge so that the staves fit smoothly into a circle, and the ends of the staves are tapered slightly in order to create the larger bilge once the staves are assembled together. That explanation probably sounds confusing, like a three-ring circus of angles, bevels and tapers, all done to curved surfaces. What's going on?

Recall, please, the concept developed by the ancient peoples: building pails and buckets with a taper allowed them to be easily secured with hoops. And then when the Celts decided to put two pails together to make a barrel, they also tapered each end. Today's stave-jointing machines have combined these multiple shapes into one step: the machines cut the proper angles on the sides of each stave in order to create a smooth circle of staves, and they narrow the stave, starting from the middle and progressing towards each end. This narrowing produces the taper, which we have found is critically important in order to make this object which we call a barrel. To accomplish this, the two blades of a jointer, set at a slight angle to each other, are relatively close together when the stave is placed into the machine. As the stave proceeds through the machine, the blades pulse – that is, they slowly separate and then close again after the midpoint of the stave. While they are cutting the taper, they are also jointing the angles on each side of the stave.

Jointing the stave edges would be relatively straightforward if all the staves were of equal width. Coopers, however, being a frugal group, make use of every defect-free stave within a 5–12-cm width range. In the jointing process, compensation must be made for these differing widths: narrower staves require only a slight angle, while a greater angle is needed for the wider ones. To put some numbers into the equation: if a cooper planned to place 30 staves in a barrel, each with the same width, the angle would be a simple 6° per side (the 360° in a circle divided by 30 staves equals 12° per stave, divided by the two sides gives 6° per side). But with staves of varying widths, and depending upon the size of the barrel, coupled with the fact that the cooper may want more or fewer than 30 staves, then the man and/or machine must compensate for these differences by adjusting the jointed angle.

In years past, preparing each stave by hand was a laborious process. The cooper would first use a heavy, long-bladed cooper's axe, a *doloire* in French, to taper the ends of the staves roughly. He would then use a draw knife to trim a convex shape on the outside of the stave and a curved-bladed adze to sculpt a bit of hollow on the inside. With the rounded and tapered stave, he would then plane each of the 'four' edges, increasing the angle for the wider staves or jointing a lesser angle for the narrow staves. The tool utilized by the hand cooper to join the staves was a 2-m-long, hardwood board, inclined downwards. A steel blade was positioned about one-third of the way along the board. The cooper would push the stave down the board, passing it over the blade to join the proper angle and being careful not to plane his fingers in the process. All these steps are now accomplished with just a few machines.

When a paper globe is made, the world map is first printed on a flat piece of paper and then wrapped around the form to complete the sphere. In a similar manner, the planed and jointed staves are first laid out flat on a table, whose length is the exact circumference of the eventual cylindrical barrel. On the table, narrow staves are interspersed with wider ones in order that the entire set of staves will bend evenly when heated. The display also provides an excellent opportunity for the cooper to make one final check for any defects before the assembly process. Also at this point, the cooper designates one wide stave into which the bunghole will later be drilled.

From the table, the barrel is raised. Picked up one at a time, the staves are placed side by side within a truss hoop, a thick, strong steel hoop which is used over and over. American coopers place the staves into a truss hoop laid on the floor, while European coopers typically use several staves to support the truss hoop at waist height, then

A French cooper is preparing to hand join a stave.

proceed to fit the remaining staves into the gaps. Regardless, the end result is a circle of straight staves spreading outward like ribs or the petals of a flower. French coopers have termed this, appropriately, *la rose*.

When crafting a wooden boat, it is necessary to heat the strakes to bend them so that they conform to the boat's curvature. Likewise, to bend the barrel's staves, heat must be applied in some form. The traditional heat source, used for the past two millennia and still employed in today's wine-barrel cooperages, is a small fire of wood scraps burned in a wire or metal basket. With the circle of staves placed over the top of the basket, the fire warms the small amount of residual moisture left within the staves. After roughly twenty minutes, heated enough to become steam, this now-hot moisture provides the flexibility to bend the staves. During the heating process, a cable which encircles the splayed ends is slowly tightened, bringing them all evenly together to form the barrel. And as with many aspects of crafting a barrel, firing is an art: apply too much heat too quickly and the inside of the barrel burns and blisters; or force the staves together before they have been sufficiently heated and they will crack on the outside. Water is sometimes again employed during the heating process:

When the staves are assembled, they are splayed outward and resemble a rose. In this photo, the cooper is about to encircle them with the cable, which will be tightened to bring the staves together while they are heated on the bending fire.

A just-assembled barrel on the bending fire, with two coopers tightening the hoops.

adding a bit of moisture to assist with the steaming or providing a cooling mist to prevent rapid heating.

Once the cable brings the staves tightly together, another truss hoop is slipped over that end. The cable is then released and other hoops are put on the barrel to hold the staves in place.

There is a second part to the firing process. This is to set the staves in their curved position – thus several truss hoops are required to keep them snugly in place while this is being done. If there is any misalignment in the jointed surfaces, the cooper will also nudge them into place using a bit of gentle persuasion from a 2-kg hammer. (This additional firing is also when the wine barrel's 'toasting' occurs, or the char in whiskey barrels, details of which will be described in the following chapters.)

However, where the 'flavour' of the barrel is not a requirement, such as with a barrel used for nails or dry goods, the final heating is rather a quick, simple process. When beer was fermented and shipped in barrels, the pressure created by the fermented brew often necessitated that the insides be coated with pitch to prevent leakage through the wood, so an interior toast level was not an issue. With barrels made from certain other woods or used for other commodities, interior coatings such as paraffin or fish glue were used to minimize leakage. Therefore this second heating was used merely to set the curve of the staves.

During the toasting, and in fact the whole barrel-building process, the cooper's 'quality assurance' programme is accomplished largely by eye, with the other senses playing minor, but important, roles. Besides looking for defects throughout the building process, during the toasting he is scanning the interior of the barrel for the proper colour of the desired toast level. If he smells a sweet aroma emanating from the fired barrels, somewhat like bread in the oven, it indicates that the toasting fires have the correct amount of fuel. A burnt odour indicates that fuel has been added too quickly, and the cooper needs to dampen the fire with a bit of water. The sense of touch is another reliable indicator: a traditional cooper uses his hand to check for the heat on the outside of the barrel, feeling for when the heat penetration is enough for the desired toast level. Cooperages now have sophisticated temperature probes and the results are recorded for each barrel in order to reproduce the exact toast level year after year. Additionally, a solid clanking noise, from the hammers hitting hoop drivers, tells the cooper that the hoops are sufficiently tight.

Once bent and toasted, the cylindrical form of the barrel now needs some ends, or heads. When the heading is sufficiently dry, it too is planed, but planed flat, with the edges jointed perpendicular to the surface. Normally, reed, or rush, is inserted between each head-board, somewhat similar to the oakum packed between the strakes of the wooden boat. The reed acts as a gasket to the heading pieces which are held together by wooden dowels or stainless steel pins. An alternative to the dowels incorporated by a few cooperages is to link the pieces of heading by finger-joints. Regardless of the connection method, the pieces of heading are placed together to form a large flat surface, with enough diameter to be cut into a circle for the head. Today's heading machinery cuts this circle, and the bevelled edge on which the staves fit, in one continuous process. Hand coopers used a traditional bow saw to cut the circumference of the head, and a draw knife to form the bevel on the edge.

The bevel of the head fits into a groove, called a croze, at the ends, or chimes, of the staves. A barrel's chimes range from blunt for barrels that expect minimal handling to a longish, chisel shape for wine barrels which are frequently handled. Traditional coopers used an adze and several differently shaped planes to chamfer the stave's chime area, giving each one the required shape – another laborious and demanding task. The croze was cut with a uniquely bladed plane. Today's croze machines have dual cutter heads (the complexity of which would make even a seasoned knife thrower blanch) that quickly and uniformly cut all the different angles and grooves in the stave ends in one pass. The circumference of the cut in the barrel is exactly that of the head, so there is no longer any need to perform what was in the past a very time-consuming task – sizing each and every head!

To install a barrel head, the cooper loosens the hoops on one end of the barrel, which allows the staves to splay open slightly. He then places the head into the groove of the croze and retightens the hoops. Barrel hoops are tightened or removed by using the force of a heavy cooper's hammer via the hoop driver or by using a mechanical hoop driver: a hydraulic machine (the modern equivalent of the 'buffalo' machine) with eight to ten flexible arms that can close around the barrel and push the hoops down with tremendous pressure.

Unlike heavy hammers used for rock or cement work, the cooper's hammer is unique. It is a small hand sledge, with one flat face used to hit the hoop driver and a long peen (the shaped end of a hammer's head, opposite the face) on the other end. This peen constitutes the major difference, as it is parallel to the handle rather than perpendicular. The advantage of this shift in peen direction is that the leading edge can be used to hit the underside of a hoop to knock it up and off the barrel.

## Hoops

Usually, it is at this stage of inserting the heads that the reusable truss hoops are replaced by what will be the finished hoops for the barrel. New hoops for wine barrels are galvanized steel, which resists the dampness of the cellars and the acidity of the wine. The chateau wine barrel, a stylized modern equivalent of the older-type barrels, will often get additional wooden hoops, made from branches or saplings of chestnut, willow or elm. These wooden hoops are an artistic and nostalgic carry-over from the period before steel hoops were in common use. However, they do have a practical side as they avoid marking the tiled surfaces of the cellar floors when the barrels are rolled. Hoops made of mild steel (low-carbon, easily worked steel) are used for bourbon barrels, primarily to minimize the overall cost of the barrels. Steel hoops, whether galvanized or not, are flared to fit the curvature of the barrel sides, accomplished by running them through heavy steel rollers which curves and flares them at the same time.

Galvanized hoops are used for wine barrels because wine is acidic and can be corrosive – the barrel might leak or wine may be accidently spilled while filling or topping up. Most galvanization holds up quite well, even after eight to ten years. However, I have seen wine corrode the galvanization if there is a constant drip. And while the galvanization generally resists wine, it does not resist the effects of sulphur, which is often burned in empty barrels to discourage the growth of vinegar bacteria or moulds. In a rather unique situation which I observed at a new, and uninitiated, winery, someone came up with the less-than-brilliant idea that instead of burning sulphur in each barrel, they would burn it in the enclosed room full of barrels, each with the bung removed. While the sulphur gas probably had the desired effect on the insides of the barrels, it had a major deleterious effect: completely stripping the galvanization off the hoops of all their new French oak barrels.

For liquid-tight barrels, the final steps are to drill the bunghole, test for leaks and do any finish work, such as sanding the outside. For slack barrels, only one head would be installed, the other being placed in the barrel after the product has been placed inside.

# TEN

# Wine: Barrels and Oak Ageing

On stave and hoop the long year through
We work'd with will and pleasure,
And when the cask was firm and true,
We press'd the vineyard's treasure.

Introductory stanza of 'A Vintage Song', fragment from the
unfinished opera *Loreley* by F. Mendelssohn-Bartholdy

 Wine was a constant companion when my wife and I were living in California's Napa Valley, as well as a topic of conversation at dinner parties. At one small soirée some years ago, when asked if I was in the wine business, I suggested to the other guest that, ah, well yes, I sell romance! There was an audible sigh of disappointment when I continued, telling her that the products were wine barrels. This prompted an additional explanation: for me, the barrels symbolize the sensual and intimate attraction of wine, a draw which promotes a closer friendship when people share a bottle at a meal.

Almost any image of barrels evokes nostalgia or a bit of the 'old country'. Wooden wine barrels, quietly resting in a cellar, suggest the essence of wine, further reinforcing the notion that this is how the best wines are made. Be it sharing a bottle at a summer picnic or at a winter's dinner with good friends, just having wine at festivities and knowing that barrels are part of the winemaking process adds an air of authenticity, fellowship and, well, romance. Perhaps this is why most premium winery brochures and websites almost always, along with a vineyard photo, include a picture of wooden barrels.

Besides adding a romantic touch, what makes a wine barrel different from a whiskey or other wooden barrel? Compared to other wooden articles, wine barrels are the equivalent of an exquisitely crafted living-room credenza or an elegant dining-room table. Whiskey barrels are more like a rough-hewn picnic table or the workbench in the garage: functional but roughly built and crudely assembled. The wine barrel is the epitome of the highest skills of the cooper: a melding of the finest oaks with superb craftsmanship. With winemakers requesting oak of the tightest grain, this means staves from the slowest-growing trees. The woods selected for those wine barrels must then be free of defects, as the cooper does not want to see his barrels leaking wine – a red pool or stain on the cellar floor. And the craftsmanship must match the quality of the wood – assembling the staves seamlessly, accurately sizing the heads to fit perfectly within the croze and fitting the hoops in precise alignment.

## The Evolution of Wine-barrel Usage

As most home-winemakers have experienced, the process rarely starts with the best equipment. Just to own a brand-new, 400-litre stainless-steel fermenting tank would be the dream of many a novice winemaker. Usually, however, and much to the chagrin of the other family members, many fledgling winemakers have utilized the household bathtub to ferment their starter wines. And far too many have attempted to age their wine in used barrels, which are vinegary in aroma, cloudy with grey-green mould and riddled with borer holes. But if the winemaker persists, he or she eventually acquires the best equipment for their particular situation. So it is most likely that it took many hundreds of years for the white oak to be selected as the preferred wood for today's wine barrels, an essential ageing tool for premium wines.

This is a process which has been evolving over thousands of years. As with many evolutionary processes, it has been influenced by several interrelated factors. One was which types of wood were readily available to the cooper – which was especially important in early years, when transportation by wagon or ship had a deciding influence on the distance the staves and heading could reasonably be moved. Another factor concerned the types of wood that were conducive to making barrels – that is, that could contain liquid, was bendable enough to make the barrel, was dimensionally stable when dried and could resist decay. And from this narrowed selection of woods, which ones offered flavours which added to, or at least did not detract from, the taste of the wine?

Pine, for example, is readily available throughout much of the world, and it is easily worked, but it imparts a resinous taste to wine and is subject to rapid decay when continually soaked with liquid, so its use for wine barrels is limited. White oak, on the other hand, has many of the desirable properties and is now considered the wood of choice, as discussed previously.

The changing use of wine cooperage in the Californian wine industry over the past 240 years provides an excellent snapshot version of the evolutionary processes at work. Economic, political, geological, geographical and cultural pressures which influenced the West Coast changes would mirror those in most of the wine-producing regions throughout the world. The time frame for these evolutionary changes in California was practically instantaneous when compared to the same processes in the wine regions of Europe. However, in the world's newer wine regions (South Africa, Chile, Argentina, Australia, New Zealand and China), the developmental stages and time frame have been similar to the California experience.

A 17th-century print by Jacques Callot depicting the activities of a grape harvest, with barrels and wooden tanks featured prominently.

In California, the Spanish missionaries were the first to make wine from the Criolla grape, brought from Spain in the 1780s. Once fermented, the storage of that wine would have been a rather haphazard affair, utilizing any available container – ceramic pots, crude wooden tanks or even animal-hide sacks. A wine critic tasting those wines by today's standards would have noted the vinegary aroma and a flavour of wet horse blanket, with raw tannins on the finish, as he gagged and rushed to spit it out. But for the padres, just having some alcohol to go with their acorn mush and mutton stew probably outweighed all the detrimental effects of substandard containers and dirty-feet winemaking. Eventually, as the missions became established, and regular ship traffic brought goods from Mexico and Spain, better wooden tanks could be imported and real barrels, which had contained water, rum or imported wines on the ocean voyage, could be utilized. The missions marched their way from San Diego north to San Francisco and Sonoma. There, they would have encountered the forests of giant coast redwoods, *Sequoia sempervirens*. The availability of those tall, straight trees, with their relatively easy-to-work attributes and minimal off-flavours, began a small industry manufacturing excellent wine-fermenting and storage tanks.

Tales of California's riches filtered out and other Europeans came to stake their claims. In 1829, a young Frenchman from the village of Cadillac and with an appropriate surname, Jean-Louis Vigne, planted vines near Los Angeles.[1] In addition to his skills as a vintner, he also had training as a cooper. He was able to utilize some of the oaks growing in the nearby San Bernardino Mountains to make the barrels required to transport his wine to customers on the East Coast of the United States. This southern California oak, while not an ideal wood for barrel making, was readily available. With the United States' transcontinental railways still a few years away, transporting the wine required a voyage by ship, a journey of several months that had to traverse Cape Horn, at the southern tip of South America. He probably shipped the wine when it had completed fermentation, and it would have aged in the barrels en route.

Grape growing and winemaking in California really exploded with the influx of gold miners during the 1840s and '50s rushes.

When the gold diggings petered out, a number of these intrepid souls turned to grape growing, as the temperate climate with its winter rains and dry summers, and the rich, volcanic soils, seemed ideal. Utilizing winemaking skills brought from Italy, France and Eastern Europe, they acquired land and built stone wineries in the Sierra foothills and coastal valleys. Wine-storage containers in those early wineries would have been, like those for the missionaries, probably anything these winemakers could lay their hands on. And although there may have been more barrels available, brought by the ships which delivered goods and products to the miners, the ideal barrels would have been the oak barrels in which whiskey, wine and ports were shipped to California from the East Coast and Europe. But others were probably made from a variety of woods, and the enterprising winemakers would have had a chance to test them all, gradually acquiring the desired wood types from the best sources. Many wineries obtained oak casks and tanks from Europe. For example, the Beringer Brothers winery, located in the Napa Valley's St Helena, purchased several German oak casks in the late 1890s. These casks, with finely carved heads, can still be seen in that winery today. Other sources of small oak barrels would have been the American whiskey industry, shipping their beverages to the miners, and the cooperages which had sprung up to meet the needs of the growing California and West Coast populations.

The term 'tank farm' has been used in the United States since the early 1920s to describe clusters of storage tanks for petroleum oil. Recently, American wine critics have coined it to aptly describe the mega-wineries where row after row of huge, gleaming, stainless-steel wine-storage tanks can be seen towering above the surrounding vineyards. The term might also have been applied to the early Californian wineries, as they typically utilized numerous tanks and vats, with few barrels. However, by the late 1800s and early 1900s, the need for barrels increased, not so much in the wineries, but, as with Vigne, for shipping the wines by rail to markets in Kansas City, Chicago and the East Coast. Cooperages sprang up to build the required casks, including the California Barrel Company in 1883 and the Carl Cooperage Company, both in San Francisco, and another in St Helena in the Napa Valley, with others in Portland and Vancouver.[2]

The wood, predominantly American white oak, was shipped from the stave mills favoured by the bourbon cooperages in the eastern United States.

Readers familiar with California's oaks, including its 'white oak', may wonder why they were not used. The wood of California's white, blue and black oaks is hard and fine-grained but brittle, making it difficult to bend without breaking. Perhaps Monsieur Vigne's cooperage made a barrel with little bilge – that is, just a slight curvature – to compensate for this difficulty, or possibly he applied some gel-type lining to the interior in order to minimize leakage if the staves did crack. The California white oak is suitable primarily as firewood, while the blue oak is used for ornamental woodworking and the black oak for flooring and some dimensional lumber uses.[3] Another West Coast oak, Oregon white oak or *Quercus garryana*, grows as far south as the San Francisco Bay area and is generally used in cabinetry. It has had some recent commercial success for wine barrels; however, the amount of quality timber is limited, especially compared with the vast forests of eastern white oak.

Besides the brittleness, another drawback of the California oak was the growing location of the trees. I noted previously that the ideal forests for cooperage wood, as well as for most lumber, are densely planted, which forces the trees to grow straight up and also minimizes lower branching. This is not the case in California's lowland forests and foothill regions, where the oaks typically grow. There, due to competition for water during the dry summers, the trees have more space between them. This space allows light in and results in lower limbs, creating numerous knots. Although growing more densely in the wetter hillside forests, the oaks there are not usually of sufficient size to make them economically viable for lumber production.

Around the world, there are other oaks and woods from which barrels have been made, but typically they do not measure up, for a variety of similar reasons, to the white oaks of Europe and the eastern United States. And despite some woods in their respective countries being suitable for generic barrels, cooperages in Australia, South Africa, Argentina and China currently import European and American wood for their wine barrels.

America's nationwide prohibition against alcohol, starting in 1920, decreased all activities surrounding the growing of grapes and making wine, including those businesses related to building wine barrels and tanks. A few wineries survived by making sacramental wines, but it wasn't until several years after the Prohibition repeal, by about 1936, that the California wine industry, and its associated cooperages, recovered to resume their normal practices. The wineries that did survive primarily utilized large tanks rather than the smaller barrels.

Fast forward to the 1960s and '70s, when several premium wineries in Napa and Sonoma counties were starting to import container loads of French oak barrels. At that time French oak barrels basked in the kudos gained from ageing many of the world's finest wines, which the winemakers of these up-and-coming California wineries were attempting to emulate. In 1976 they succeeded, or at least imagined that they had, based on the famous Judgement of Paris tasting in which the Napa Valley's Chateau Montelena Chardonnay was chosen number one, by a panel of French judges, over a number of prestigious French wines. With wine of this quality, impressive enough to command the prices necessary to continue to purchase and use French oak barrels, those barrels became all the rage.

In California's premium wineries, French oak barrels have been in continual demand ever since. However, not all wines can command a price to justify their use: much wine is made to be *vin ordinaire*. The evolution of wine in California also includes far less expensive wines, utilizing less expensive containers. For these other wines made in barrels, American and Eastern European oak, and oak alternatives – other methods of adding oak flavouring to wine, such as toasted oak chips or slats – are currently important substitutes.

## Using Wine Barrels

How does a cellar rat, as a winery's cellar staff are affectionately known, get the wine in and out of the barrel? For most wine barrels, the only orifice is the bunghole. The cellar rats use a hose placed in this 50-mm-diameter opening to pump or pressurize wine into or out of the barrel. When the bunghole is drilled at the cooperage, it is then cauterized with a hot, tapered branding iron to assist in sealing the bung. Historically, a wooden bung was used to stopper the bung-hole, keeping air and dust out of the wine. Today, it is more common to see silicone bungs in use.

As quality silicone became available, winemakers recognized its advantages over wood for bungs. Those made of silicone are impervious to the wine, cleanable and, with their flexibility, can be pushed in hard to make an airtight seal. Wooden bungs needed to be covered with wax or sealant to prevent wicking and to make them cleanable. And if the wooden bungs were not inserted in the same orientation each time, they deformed, requiring hammering in harder to obtain a tight seal, which led to further deformation. Most bungs were made of a softer wood than oak and were thought of as rather expendable when it came to hammering them, rather than also deforming the bunghole and its stave in the process. However, some hard-headed winemakers insisted upon using hard-oak bungs, hammering of which deformed the bung stave as well as the bung, weakening and sometimes breaking the former.

During barrel fermentation of chardonnay, glass bungs are occasionally used. Their seal is just loose enough to allow the fermentation gases to escape. Perched atop the barrels in a *chai*, the French word for the cellar where barrels are stored, they look exquisite, especially when lit by candlelight – and with the soft, bubbling sound of fermentation in the background.

When wine is ageing in the barrel, it will usually be racked one or more times during the maturation period. This process separates the clear wine from that still containing some residual sediment from the fermentation. The normal method for racking is to draw off the clear wine via a hose inserted in the bunghole. This can be accomplished with minimal aeration to the wine. However, when additional wine aeration is needed, such as for heavier, Bordeaux-style wines, some wineries utilize an esquive: one of two small holes in one of the heads of a barrel. With the barrel resting on its side and the main bung facing straight up, one esquive is near the bottom of the head and the other is slightly above and off to one side. These are used after the wine has had several months to settle. The winemaker or cellar rat will remove the upper esquive bung to drain off the clear

Two cellar workers are shown racking wine through an esquive
in the head of a barrel. To facilitate the racking, this esquive
has a valve instead of the traditional wooden bung.

wine, splashing and aerating it into a copper or stainless-steel bucket.
Removing the lower bung then drains wine with some sediment,
which is collected for further clarification. The wine below the lower
esquive is usually full of sediment and is discarded.

While the shape of barrels has been largely unchanged for 2,000
years, their outer appearance is now cleanly buffed, adapting to
the twenty-first century's renewed interest in wine and the winemaking process, including winery tours. Fifty years ago, most wine barrels
were at least planed on the outside, but few were sanded smooth as
well. Now it would be unusual to see a barrel not cleanly sanded, and
most are ultra-sanded to the point where they are absolutely pristine,
befitting the exquisite cellars and chateaux in which they rest. Because
the barrels are now so 'clean' in terms of minor blemishes in the
wood, coopers occasionally get frantic phone calls from some winemaker customers on receiving their new barrels. The winemaker will
question whether the tiny knots or the darker-colour mineral streaks
that do appear might leak. Years ago these would never have been

These chateau-style barrels, ageing wine in a wine *chai*, have thinner staves and headboards, requiring the horizon brace on the heads to withstand the pressure. The space between the centre hoops has been stained red to hide any spills.

noticed; and, no, with the testing and quality control carried out by modern cooperages, those surface blemishes rarely leak.

The trend towards elegance in the cellar extends to wine stains. The French have come up with a most practical solution: they paint the area around the bung and between the two middle hoops with a red stain made from grape pumice. This burgundy-coloured band tends to camouflage all spills. Additionally, the way in which this band contrasts with the tan and blond wood of the barrels and the dark brown of the wooden hoops traditionally placed on some wine barrels lends a striking look to these barrels.

What started as a simple way to identify which cooper made the barrel – each would place his individual mark or sign his name on the head of the barrel – has grown into full-blown marketing for both the cooperage brand and their winery customers. As the cooperages developed, marketing-savvy owners would also brand or paint the name of the cooperage on the barrel's head. Really marketing-savvy owners put their name on *both* heads, so no matter which way the barrel is positioned in the cellar, their logo is promi-nent. Today's cooperages utilize modern laser printers to etch the cooperage logo and information about the barrel – such as the source of the oak, the toast level and when the barrel was made – and also to add quality-control details. Laser printers can also easily imprint the name and logo of the winery and/or particulars such as harvest date, type of wine, storage location, barrel code and so on – more useful marketing tools.

Loggers, millers and coopers have gone to all this effort to make a functional work of art, so how long does it last? For wine barrels, the limits to their 'use-by date' are determined by taste (oak flavouring) and sanitation. The oak flavours subside to near zero after four to eight years, depending upon the alcohol levels of the wines ageing in them. Wineries seeking that oak influence in their wines must renew all or a portion of their cooperage annually. Once depleted of their oak flavours, the neutral barrels will continue to provide ageing for a few years longer. However, for wine barrels that are in use for much longer than ten years, the potential for wine spoilage increases as microorganisms gain a foothold within the porous wood. Wineries go to great lengths to sanitize barrels, keep them full and otherwise ensure against the wine going off within them. However, with age, and the repeated opening of the bung to fill or empty a barrel, the potential for airborne bacteria and moulds to enter the barrel is inevitable. Thus, in our local garden centres, we see numerous otherwise fine-looking wine barrels cut in half for planters or beds for pets after serving for a maximum of only eight or ten years within a winery.

Because of the higher alcohol levels, barrels for spirits can be used for considerably longer. I have seen barrels in Scotland that have been in use for over 60 years. With care – involving stave and hoop replacement – they continue to provide the ageing that whisky requires.

The unfortunate fact is that less and less wine is being made in wooden barrels for the simple reason that barrels, especially those of French oak, are becoming too expensive for their use to be justified in the generic winemaking process. Making wine in barrels is costly. Besides the annual expense of the new barrels, there is the extra cost of the time required for ageing. And, while forklifts, metal barrel-pallets and automated cleaning and filling lines have speeded up barrel processing, it is still laborious. Each barrel must be filled and then periodically topped up, as the water within the wine evaporates from the barrel. Several months into the ageing cycle the wine must be racked, with the clear juice placed in a clean barrel, and then the original barrel cleaned and refilled. This process is often repeated three to four times during the course of a two-year ageing regime. When the barrels are empty, they must be maintained with a sanitizing agent such as sulphur gas, to prevent the growth of moulds and bacteria. For wineries with hundreds or thousands of barrels, processing the wine in barrels is a full-time job for several cellar rats.

But where the cost of purchasing and using barrels can be justified economically by the price of the wine, barrels do contribute to enhanced flavour, complexity, colour and stability. Barrel ageing primarily benefits wines such as Chardonnay, Sauvignon Blanc, Pinot Noir, Cabernet Sauvignon, Merlot, Cabernet Franc, Gamay, Syrah, Zinfandel, Malbec, Grenache, Sangiovese, Nebbiolo and Tempranillo. Of course, a number of other varieties and styles of red and some whites are also enhanced by barrel ageing. And some wines, such as Gewürztraminer, are tasty without being placed in small barrels for ageing.

Woodcut of a barrel with a funnel for pouring in the wine, 16th century. Note the multiple wooden hoops encircling the barrel.

For many premium wineries, the annual price of cooperage is second only to that of grapes. As the use of small barrels took off in California, in 1982 the author of an article about ageing wine in wood was astounded that the price of American oak casks was then $100 and that of French oak barrels over $300 each. Today, the price for the least expensive American oak barrel is $300, and the French oak ones hover around $1,500 each! With these price increases, which wineries can afford to make wine in barrels? And what factors account for the expense of the barrels?

The standard 225-litre wine barrel stores enough wine to fill about 300 standard 750-ml bottles (about 25 cases). Barrels are typically used for three vintages, ageing the wine for 18–24 months for each vintage. During that period, the barrel will have aged 900 bottles' worth of wine. To simplify the example, let's say that the barrel cost $900 (actually this price is about midway between the barrels made from American oak, in the $300 to $600 range, and those of French oak, in the $1,000 to $1,500 range). Amortizing this cost would charge $1 to each bottle produced, which is not much for a wine wholesaling at $20, but is a huge percentage for one wholesaling at $3. Needless to say, it is generally only higher-priced wines that are aged in small oak barrels.

However, barrels come in a range of prices, so it is possible for wineries with huge volumes to utilize barrels for some ageing and still sell the wine at reasonably low prices. Further savings may be gained by using American oak barrels, which are around half to one-third the cost of French oak barrels, or barrels made from Eastern European oaks, which tend to be about midway between the two. I noted in chapter Eight the significant differences, and the resulting cost effects, between making staves from European oak versus American oak. Additionally, while American deciduous forests are not quite as large as those in Europe, they are, by and large, privately owned. The sale of logs is normally made directly to the mills (as opposed to being purchased at auctions), which helps to keep the initial timber costs lower. Yield advantages from processing methods plus lower American labour costs also help minimize the costs.

## TCA: Caution with Barrels

In the late 1980s, some American wine critics began to suspect that certain highly touted wines were really not all that good; they actually had some problematic tastes and aromas. The flavour of 'mouldy cardboard' or the smell of 'wet horse blanket' were some of the terms used to describe the wines. These phrases had routinely been applied to *vin ordinaire*, but not to the top-growth chateaux and high-end wineries. Initial diagnosis pinned the cause on the corks used to stopper the bottles, and the resulting term was that the wine was 'corked'. Subsequently, through extensive research, these off flavours and aromas were shown to be caused by a compound known as TCA (2,4,6-trichloroanisole), along with several other similar compounds. Moulds found naturally within the cork, combined with the chlorine of the cork-cleaning and bleaching process, and in the presence of certain hydrocarbon precursors, formed the TCA.[4] However, by the early 2000s further research had ascertained that it was not only corks that were found to produce TCA, but barrels as well!

According to James Laube's 2002 article in *The Wine Spectator*, the wines from one well-known Napa Valley winery were suspected of TCA.[5] A subsequent investigation by the winery revealed that the chlorine originated from city water that was being sprayed onto barrels as micro-droplets through a cellar humidifier. As the cellar workers opened the barrel bungs for topping or cleaning, the minuscule chlorine droplets floating around in the air would settle into the wine, mix with the naturally occurring moulds in the barrel and cause TCA in the wine. The TCA-tainted corks may have only exacerbated the problem by the time the wine got to the consumer.

At first, cork companies were pilloried. The can-do spirit of the Australian and New Zealand wine industries went so far as to shift the majority of their bottle closures from corks to screw-caps. A deluge of synthetic corks were introduced into the market to replace natural cork. But with further understanding of the chemical processes which produced TCA, the entire wine industry, including the cooperage companies, started looking in all the corners. As the manager of a wine-barrel cooperage, upon instituting an inspection, I realized that our company was also using city water. We set up an elaborate filtration system to remove the chlorine from the water which we used to wet and test the barrels. Other companies shifted to non-chlorinated well water. And cardboard, with its chlorine-bleached material, was minimized in the barrel packaging.

The upside of this activity has been that, while corked wines still occasionally appear, the incidence has been dramatically reduced. Most wineries are now much more cautious in their use of any chemicals. And the entire process of making the wine barrels has been scrutinized, from obtaining the wood in the forest all the way through to delivering the finished barrels to the winery. Any place or process where the wood or the barrels can possibly interact with potentially damaging chemicals – chlorine or others – has been changed or eliminated. A number of cooperages now certify their sanitation regimes.

Despite all this effort, on a recent tour of the Hunter Valley wine region north of Sydney, Australia, I found a strange taste in a red wine at one winery. When I tasted the same off flavour in all the winery's other red wines, I was certain they were 'corked' and that the culprit was TCA. The wine bottles had screw-caps, so corks were not an issue. The tasting was followed by a tour of the winery. On the tour, our winemaker guide indicated that only the red wines were being aged in barrels. Since corks were not being used, the wine barrels were probably the culprit. But it was not just one brand of barrel: the whole cellar was suspect, as the barrels were from a mix of cooperage producers, and all the reds were 'corked'. A look into the barrel cellar revealed that cardboard cases were being stored alongside the barrels. I concluded that it was possible that the humidity within the cellar was atomizing chlorine contained within the cardboard from its processing, which then got into the barrels. Equally, if the winery was using some barrel-cleaning treatment which involved chlorine, that could also be a cause. I suggested both of these possibilities to the winemaker in the hopes that he would correct the problem.

## MicroOx: Mimicking the Barrel

One year, while travelling through France, our companions, one of whom was a Napa Valley winemaker, slipped off for two days to the

Madiran region of southern France. At the time, it seemed like a secret mission – spying on the competition or some such nonsense.

When they returned, I realized that it was not nonsense, but it did involve a bit of industrial reconnaissance. The Madiran vineyards, lying southwest of the Armagnac region, grow the Tannat grapes which, as the name implies, produce a dark red, tannic wine.[6] That heavier style may have been an acceptable taste profile years ago, when wine consumers were expected to cellar their own wines, providing the extended ageing and its attendant maturation and softening. However, with today's pick-me-off-the-shelf-and-drink-tonight mentality, the vintners of Madiran knew they needed a lighter, softer style to be competitive within the wine marketplace. Years in barrels would soften the wine, but economic pressures within the industry could not justify the added costs of extended barrel ageing. Their dilemma was how to replicate the barrel's ageing and mellowing processes at a minimal cost and in a reduced time frame.

Instead of placing the wine in barrels to interact with oxygen, they reversed the process and brought air to the wine. They did this by introducing oxygen, via a diffuser, while the wine was in the tanks. As the oxygen was slowly pushed through the diffuser, they carefully monitored the increase in dissolved oxygen. This process mellowed and softened the Tannat wines within a relatively short time period. The technique is now termed micro-oxygenation, or microOx for short. My winemaker friend visited Madiran to learn about this technique's progress and taste the wines for himself. He came away reasonably impressed, and now he and many other winemakers routinely utilize the microOx method.[7]

Subsequently, since the mid-1990s, the use of microOx has exploded worldwide as winemakers see its advantages. It has been employed not only to shorten the ageing cycle for supermarket wines, but also to address specific problems – such as reducing high pyrazine (the 'bell pepper' effect), a 'sandy or grippy' tannic mouthfeel or elevated hydrogen sulphide – in both red and white wines.

At some wineries, microOx is replacing barrels: they are 'ageing' the wine in large stainless-steel tanks using microOx and adding oak flavouring by introducing toasted-oak chips or staves. Many low-priced wines are made using this recipe. At other wineries, it is just one more resource in the winemaker's toolbox and, like barrels, has become an extremely important one.

## Oxygen and Oak: Barrel Maturation

In her book *Wine Uncorked*, Fiona Beckett discussed barrels, noting that: 'What oak does contribute, however, are more interesting, complex flavours and the ability – particularly important in fine reds – to age.'[8]

The earliest winemakers recognized that wine undergoes dramatic change, and ages faster, in small barrels as opposed to larger tanks. And as both the art and science of winemaking and barrel production progressed to a point of some consistency, and if the winemakers had an opportunity to try barrels from different coopers or woods from different regions, they noticed that the wine was enhanced in diverse ways – the tastes developing in barrels from one cooperage might enhance the wine better than in barrels from another cooperage. This could be a result of various combinations of the wood itself, its origin, the drying process and the toasting regime, as all contribute to the tastes imparted by the barrels.

Astute winemakers request a certain barrel brand and toast level or purchase combinations of brands to add nuanced complexity to their various wine varietals. While there have always been a few winemakers who have carefully crafted their wines utilizing small cooperage, serious interest in the fine distinctions of oak wine barrels got into full swing in the 1970s and '80s, as the making of premium wine exploded worldwide. And with that interest came a demand for answers: what are the flavours imparted by the barrels, and how are these flavours conveyed to the wine? Research scientists at universities with specialized wine departments in America, France, Australia, South Africa and Germany, along with oenologists in the largest wineries such as Gallo or on retainer for some of the major cooperage houses, answered some of the questions, but also opened areas for further investigation.

## Oxygen in Wine

If you have ever tasted an extremely young wine, still fermenting or having just finished, you probably noticed how sharp, rough and tannic it was. Did it make your mouth pucker? To soften the flavours, as mentioned above, the wine is often placed in barrels for ageing. For a simple explanation of how barrels 'age' wine, and change these youthful characteristics into the supple, rounded mouthfeel of a matured wine, I remember the words of a perceptive tour guide on one of my first winery tours:

> Imagine a molecule of young wine as a three-dimensional star with many sharp points. These sharp points produce the scratchiness and roughness in your mouth. The barrels are used to 'age' the wine, and it is during this 'ageing' that the sharp points of the wine's 'star' molecules are smoothed, leaving rounded, softer edges, which then equates to a mellowed and softer mouthfeel.

This is a great analogy to describe what happens, but the actual chemical processes are extremely complex, involving the inter-action of literally hundreds of compounds. And while those pro-cesses may thrill the heart of a biochemist, here I will just provide an overview, emphasizing the barrel's role in this intriguing molecular dance.

Wine in the barrel evolves, and the oenologist, or winemaker, must constantly monitor its changes. He or she is observing two main reactions: the oxidation, which is the combining of minute amounts of oxygen with the various wine molecules; and the incorporation of the raw and toasted oak flavours into the wine.[9]

Let's first examine the impact of air on wine. Oxygen in small amounts is beneficial to wine, aiding the ageing process and com-bining with various compounds to increase mellowness. However, like many things, too much oxygen is detrimental, as anyone who has left wine in an open bottle for too long has experienced. Prolonged exposure to the air will oxidize the wine, or turn it into vinegar, due to the unseen hordes of vinegar-producing bacteria spores floating in the air all around us. Any time we open a wine bottle, or barrel for that matter, those spores can drop into the wine. A low level of sulphur in the wine usually keeps them at bay, but over time, and with enough air introducing more and more bacteria, they will eventually turn the wine into vinegar.

When winemaking, some oxygen is introduced incidentally during the transferring, filling, racking and topping processes. This is usually enough for most wines. For rich, heavier wines, further aeration can be accomplished by splashing the wine in buckets as it empties from the esquive bungholes.

Wooden barrels also allow oxygen to interact with wine – this is part of their uniqueness. For hundreds of years, winemakers have wondered how this happens. On removing a tightly sealed bung, they would find the level of wine lower and a vacuum in the resulting space. The lower wine level indicated that wine was being lost. A little further observation indicated that water, in the form of vapour, had escaped and that the percentage of alcohol in the wine was greater. However, the vacuum indicated that air – oxygen – had not entered, and yet aeration – ageing – seemed to have occurred! It wasn't until extensive research into this question was undertaken in the 1970s that the answer was finally revealed. It was found that within the interior of the staves and heading, a minute amount of oxygen, entering from outside the barrel, interacts with the wine that has penetrated into the wood.[10] The complete chemical, mechanical and microbiological processes are much more complex, but this explanation provides a good, practical understanding of why the wooden barrel is so unique and crucial to ageing fine wine and spirits.

'Phenolics' is a general name for the many compounds present in grapes which create aromas and flavours in wine. During the fermentation and ageing processes, it is these compounds which interact with the oxygen, combining their short, 'spiky' molecules into long, 'soft' ones. Tannins are another group of compounds which originate in both the grape and oak wood and in some cases cause the astringency, or pucker-factor, noted in young wines. From tannins' interaction with the minute amounts of introduced oxygen, they too become linked, forming larger molecules which are softer on the palate.[11]

Colour in wine is due to the anthrocyans: the red/purple pigments. Oxygen causes these pigments to intensify, adding and strengthening the colour in red wines.[12] Additionally, oxygen aids in the formation of anthrocyan-tannin complexes (again, a longer, softer molecule), which in turn tone down the tannins and remove some of the astringency of young wines, akin to what the tour guide described as 'rounding of the star points'.

## Wood Flavours in Wine

Besides oxygen changing the taste of wine, as we've seen earlier, the oak itself contributes flavours.[13] Oak lactones provide coconut and vanilla aromas. Interestingly, despite the higher levels of these particular extractives in European oaks, the same flavours coming from American oak are more pronounced.[14] Perhaps this is because they are typically introduced into the wine in American oak barrels more rapidly than for those of French oak – the rapidity layering the flavours instead of allowing for a subtle integration. Because the use of American oak for ageing wine is in its relative infancy compared to French oak, perhaps just longer air-drying and slower oak toasting can mitigate these factors.

The tannins in oak come from lignin within the wood's cell walls and, being soluble in wine, can produce astringent flavours. Bitterness in wine can be attributed to tannins from the wood, acid and tannins from the grapes or some combination of these, although a limited addition of tannins can add body and help to preserve a wine. Good viticulture, good winemaking and good barrel making, individually and in combination, all seek to find that correct balance of tannins, where they are structurally integrated and balanced within the wine and contribute to an enhanced flavour and complexity.

Earlier, we noted briefly how wine also picks up flavours from the toasting of the interior of the barrel, and the oak dial was developed to show the relationship between the levels of toasting and the resulting flavours. This wheel indicates that, during the initial toasting of the wood, spicy/clove flavours are developed from the wood sugars. Continued toasting brings out the coconut and vanilla flavours and then moves into the toffee/coffee. If the barrel is left on a fire for an hour

The oak dial is a graphic image of the transition of the flavours as the oak barrel wood is heated.

MEDIUM

LIGHT

HEAVY

SAWDUST
COCONUT
VANILLA
CLOVE
WOODY
TOFFEE
SPICY
SCENTED
TOASTY
COFFEE
ALTERATION
SAPPY   BY OAK WOOD
SMOKE
DULL
BUTTER

TOAST

OAK DIAL

or more, some charring occurs, resulting in butterscotch and smoky flavours. All these flavours are more or less desirable depending upon the wine and wine style. In response to today's winemakers wanting consistency and control in order to minimize the differences between vintages, coopers have learned how to dependably reproduce these flavours in the toasted barrels. Through careful monitoring of the moisture in the oak, using temperature probes during the toasting process, applying a consistent amount of heat by controlling the amount of fuel to the toasting fires and by monitoring and recording the amount of heat each barrel receives, the odds in favour of the barrel purchased this year being similar to the one bought last year have improved dramatically.

Winemakers are very aware of these barrel characteristics and request certain toasts for certain wines. For some Chardonnay, barrels with a heavy toast may contribute a butterscotch (toffee to coffee) flavour that becomes a subtle nuance in the wine. Certain Pinot Noirs can also accept a slightly smoky character from a heavily toasted barrel. Cabernet Sauvignon and Syrah typically taste better with the spicy or vanilla flavours of light and medium toasts.

## Barrel Fermentation

Some modern wine lovers, including myself, follow the ABC rule of wine drinking – Anything But Cabernet or Anything But Chardonnay. Others believe that Cabernet Sauvignon is the king of wines and Chardonnay the queen. Cabernet has reached the pinnacle of its success primarily through the marketing efforts of Bordeaux vintners. Chardonnay, on the other hand, has been assisted by wooden barrels: it is one of the few white wines to be barrel-aged and of even fewer that are fermented in the barrel.

There are several excellent reasons for fermenting Chardonnay in oak barrels: barrels can moderate the heat of fermentation, resulting in a cooler, slower and ultimately more flavourful fermentation; flavours are extracted from the oak and toast of the barrel to enhance the butterscotch and vanilla characteristics found as part of the traditional Chardonnay grape-flavour profile; and the stirring of the lees, the remains of the fruit solids and yeasts, helps to integrate and intensify the Chardonnay fruit flavours. Stirring is accomplished far more easily in small barrels than in larger tanks.

However, in barrels, the Chardonnay can easily pick up a 'toasty' oak characteristic, and too much of that flavour may overwhelm the wine. In the opinion of many wine consumers, this occurred with a number of the California Chardonnays during the late 1980s and early '90s. Eventually, the wine industry got the message, with the pendulum swinging so far in the opposite direction that there are now 'unoaked' Chardonnays on the supermarket shelves. However, most Chardonnay winemakers have taken a more balanced approach, offering their wines with an integrated flavour profile rather than one dominated by oak.

Sauvignon Blanc is another white wine that is aged, and occasionally fermented, in oak barrels, both of which tend to tone down the grassiness. Several French wine regions, Sancerre, Pouilly-sur-Loire and, in Bordeaux, the Graves and the Sauternes, among others, are well known for delicious, barrel-aged Sauvignon Blancs. As another example, in 1968, the Napa Valley winery owner and marketing guru extraordinaire Robert Mondavi produced a dry, barrel-aged Sauvignon Blanc. He further enhanced the wine's image by relabelling it Fumé Blanc, a descriptive denoting the taste added by the barrel's toastiness.

## Wine-barrel Design and Styles

With over 100 wine-barrel cooperages worldwide, ranging from the huge World Cooperage, which produces upwards of 100,000 wine barrels per year, to small, single-person craft shops, the range of barrel sizes, wood sources, production techniques and toast levels is enormous. We have already touched on many differences, but an elaboration of the significant visual styles is also in order.

The French, *bien sûr*, developed the two most commonly used wine-barrel styles: the 225-litre Bordeaux export and the 228-litre Burgundian export. The term 'export' is derived from the French practice of shipping wine – in the barrel – from the chateau or winery where it was made to a distribution point or end-user. The export barrels were strong enough to be rolled, handled and reused: perhaps shipped to Bercy, the wine entrepôt section of Paris, and returned to the wineries. In the days when wine was shipped in the barrel, export barrels were often used over and over again like beer kegs. However,

Barrels differ in size, style and hoop configuration. This chateau-style barrel, with wooden hooping at the ends, has dimensions similar to a Bordeaux barrel and is taller than a Burgundian barrel.

as with beer, this practice has been largely superseded by shipping and delivering wine in glass bottles or tanks.

Under the export name, the two barrel styles – Bordeaux and Burgundy – evolved within those regional cooperages to meet the specialized needs of their respective wines and wineries. In particular, the Chardonnay and Pinot Noir of Burgundy are enhanced by a heavily toasted barrel. To produce this intense toast level, the barrels are on the bending and toasting fires for close to an hour or more. As a result, the staves are bent to a good curvature, producing a short, rather squat barrel of 228 litres and 90 centimetres in height which is called la pièce. It is not clear which came first: the long toasting required for obtaining a severe curvature in the staves or the appeal for the heavier toast in those wines. Most likely these two characteristics evolved simultaneously.

The classic flavour profile for the Cabernet Sauvignon of the Bordeaux region does not require a heavily toasted barrel. As a result, the barrique barrels developed by the cooperages in that region are typically lighter in toast, and with straighter sides, resulting in slightly taller barrels by 5 centimetres, containing 225 litres.

Since wine barrels made in American cooperages basically evolved from uncharred whiskey barrels, the style emulated neither the Bordeaux nor Burgundy designs. Over the ensuing years, American barrels (and American wines) at first tried to imitate French ones, but have now branched out to find their own unique styles and tastes.

Both these export styles, the Bordeaux and Burgundy, come in other sizes: half barrels (112 litres), for holding that last bit of wine for topping, as well as larger sizes (300, 500 and 600 litres). The 300-litre sizes are popular in Australia, and the larger puncheons are used when winemakers want a more nuanced oak flavour in their wine. Uniquely, a barrel of 265 litres was built at the request of the huge Gallo winery. They wanted a barrel which would maximize the volume of barrels used on the forkliftable barrel pallets already in use within their enormous winery warehouses. The resulting design combined the width of a Burgundian barrel with the length of the Bordeaux style. Seeing the success of these barrels, and wanting to maximize their own barrel storage, many other large wineries have subsequently ordered that style.

A contrast to the export-style barrels, and the epitome of wine-barrel cooperage, is the chateau-style barrel. These barrels have wooden hoops in combination with narrow galvanized steel hoops. The staves are a bit thinner: 22 millimetres as compared to 27 millimetres thick for the standard export barrel. With the use of thinner head boards as well, a brace – a wooden piece, called *la barre* in French – is placed perpendicular to the heading pieces of each individual barrel's head in order to provide additional strength. The thinner staves and heading must be flawless and each stave must align perfectly with its neighbours. These chateau barrels seem to help the wine age more quickly, but only by a month or two.

Chateau barrels are beautiful, with equal-width staves of an identical off-white colour contrasted with the dark brown, still-attached bark of the chestnut hoops and perfectly proportioned dimensions. Beside their exceptional winemaking attributes, lying on their sides in neat rows in a stone *chai*, they create a stunning and romantic ambiance for any winery which can afford to utilize them.

## Barrel Tasting

Winemakers, as we've mentioned, select their barrels based upon the flavours – spicy, fruity, toasty, smoky and so on – that can be imparted to the wine during the barrel ageing period. Despite the often-apparent similarity in the processes of wood ageing and barrel toasting between the various cooperage houses, the actual nuanced flavours of the different brands are usually discrete enough to persuade winemakers to purchase from different cooperages. To evaluate the flavours, winemakers must constantly taste the wines as they age and evolve while in the barrels. They also often allow barrel salespeople to join them in their assessments as a way for the latter to get feedback on the taste of the wines and to learn which factors in the barrel-construction process contribute to the various barrel flavours.

To deviate slightly, good winemakers must be given much credit. Those individuals who can understand, perhaps 'juggle' is a better word, all the nuances of wine maturation – the length of time the wine is in the barrel, the changes in the wine and the impact of the barrel flavours – are to be much appreciated. It is a multi-dimensional puzzle – with time going in one direction, wine changes in another and the influence of the barrel attributes going in a third – which they must solve for each lot of fermented grape juice in order to produce exceptional wines. So, to all winemakers, I extend my utmost respect for a most difficult job.

On the surface, barrel tasting is a rather simple procedure, but one with important results. It sometimes occurs in the wine lab, but is far more interesting when conducted in the barrel cellar. With tasting glass in hand, one follows the winemaker into the cellar, where he or she uses a wine thief to extract wine from the barrels. The thief is a round glass or plastic tube with small holes at either end. Inserted into the bunghole, it fills with wine. The winemaker then places their thumb over the top hole to retain the wine in the thief, extracts it from the bunghole and releases the thumb to fill the glasses.

When you expect to taste a number of wines it is far better to use your nose and eyes more than your mouth. Even with spitting into a convenient gutter, tasting wines, especially young ones, from ten to twenty barrels can be a bit overwhelming.

After swirling, smelling, tasting and spitting, a comment is required – and not 'It tastes good.' It definitely has to be something more intelligent, such as: 'This wine displays fresh raspberry, dark plum and floral characters, while the fine-grained tannins go nicely on the mid-palate' or: 'The ripe plum and chocolate toasty undertones of the barrel are complementing the spicy fruit.' Wine geek-speak, to be sure.

Geek-speak it may be, but it is part of an important process for both the winemakers and barrel salespeople. Through the exchange of flavour-profile information about the wine and the individual cooperage brands, both groups are able to make informed decisions regarding improvements; the winemakers can decide which brand, oak source and toast level goes best for each wine that they produce, and the barrel salespeople obtain knowledge about their own cooperage, compared to the competition, with which they can tweak their manufacturing processes.

If you harbour a bit of scepticism about how there can be enough differences between the various cooperage brands, you are not entirely alone. While on a trip through France, I questioned this myself. My

Winery personnel use a wine thief to extract wine from barrels for sampling.

wife and I had the opportunity to taste at Château Lafite and Château Mouton Rothschild, two of the top Bordeaux wineries. The winemakers first showed us their gleaming, white-painted, wooden tanks where the grapes are fermented and then took us into their cool, cavernous chais, filled with hundreds of new French oak chateau-style barrels.

Both winemakers produced the requisite wine thieves to extract the deep purple liquid from a barrel and handed each of us a glass of the current vintage. In this case, I think that we got away with simple comments such as '*délicieux*' or '*très agréable*' in our elementary French. However, I did ask each the origin of the French oak that they requested for their barrels. One stated that he preferred wood from the Limousin forest, which is a large timber track in the centre of France, around the city of Limoges. The other indicated a preference for oak from Nevers, in the region of the same name, northeast of the Limousin forest and southeast of Paris.

How could two equally distinguished winemakers, crafting some of the most prized wines in the world with the same grape varieties, age their wine in two distinct types of oaks? It only dawned on me some years later, after numerous other tastings: each barrel brand and style contributes a nuance of difference which helps to make the wine unique. And it is that uniqueness which creates the brand. The trick is getting the correct cooperage brand to enhance the style of wine the winemaker is trying to achieve.

Wooden wine barrels play an extremely important role in making the best wines. And while we know considerably more than the Romans and Celts, there is still much to learn.

# ELEVEN

# Craftsmen: The Coopers

*O tintamarre plaisant*
*Et doucement résonnant*
*Des tonneaux que l'on relie!*

Oh pleasant uproar
And sweet resonance
When hooping casks!

Olivier Basselin, fifteenth-century cooper
and poet, 'Poème à la gloire du tonnelier'
(Poem to the Glory of the Cooper)

## Cooper in Name Only

The MINI Cooper, that smart little sports car you see zipping around towns today, originated as a concept design car for a British racing mechanic. The Cooper Car Company, owned by John Cooper and his father Charles, chose the British Motor Corporation's Morris Mini-Minor model, with its front-wheel drive, light weight and low centre of gravity to beef up for English hill-climbing rallies. Years later, piggy-backing on the Mini-Cooper's racing success, the Mini was purchased by BMW, who changed the name to MINI Cooper and turned it into an extremely hip car that appeals to ageing Baby Boomers and well-heeled Generation Xers.

Another famous 'cooper' was James Fenimore Cooper, the American author of *The Last of the Mohicans*, among other books. During the early 1800s, he lived in Cooperstown, New York, where his father, Judge William Cooper, had purchased land and established a community. The town is now better known as home to America's National Baseball Hall of Fame. The closest either of the Coopers came to being actual coopers was that Judge Cooper may have been a wheelwright in his early days, as he had wooden wheel-making planes and shaves among his carpenters' tools when he commenced constructing the town's buildings. However, despite their distance from actually constructing barrels, most likely both these families, and other families with the 'Cooper' surname, could

Several of the barrels on the left side of this 19th-century photo have
not been fired to curve their staves. The cooper on the left is shown
with his hammer and hoop driver. The one in the centre is planing the
croze to accept a head, and one on the right, an apprentice, is shaving
the outside to clean the barrel, a finishing procedure.

trace their lineage back to someone who was an actual cooper, and
it is these craftsmen to which this chapter is dedicated.

Sadly, over the past 75 years, most of the cooper craftsmen have
gone the way of TV-repair people, among others – unemployed,
retired or forced to find jobs in other industries. The doors closed at
hundreds of cooperages throughout the United States and Europe
which at one time made barrels for beer, oil, flour, fish and a multitude of other commodities, and thousands of coopers lost their jobs.
Today, on both sides of the Atlantic, relatively few are employed by
wine- and whiskey-barrel cooperages and a bare handful for deco-
rative kegs and wine and water tanks. And while there has been an
increase in the demand for premium wine barrels, improvements in
machinery have slowly winnowed down the commensurate need for
coopers.

## Craftsman and Artist

In this era of celebrity movie stars, celebrity chefs and even celebrity winemakers, it is important to remember that they are all supported by dedicated people without whose sacrifice and expertise the celeb's rise to fame would be elusive. For winemakers, high-profile or not, the talent of the viticulturists and their crews, carefully tending the vines, only begins the process. Winery personnel – the cellar workers and sales staff – and the many skilled persons who provide the products, services and technology used in wineries also contribute to the winemaker's success. However, after the grapes, few products and people are as important to premium wine as the wooden barrels in which the wine is aged and the coopers who craft them.

The demand for wine barrels to age and flavour premium wine has actually increased worldwide over the past 50 years as more and more consumers have learned to enjoy this appealing and complex beverage. Wine-barrel orders during recent vintages have reached several hundred thousand casks sold annually to wineries around the world. However, these numbers are almost insignificant set against the historic use of generic wooden barrels and the cooperage trade now numbers only a few hundred men (and a small number of women), working in fewer than 300 cooperages and barrel-repair shops worldwide.

That said, with this recent, increased demand for wine barrels, particularly in California, several French cooperage houses have set up satellite facilities in Napa, Sonoma and Mendocino counties to satisfy their winery customers in these important wine regions. In these Californian cooperages, at least one or two coopers who understand the complete barrel-fabrication process are required. These 'Master Coopers' haven't necessarily migrated from Europe, but most have been trained in Europe, particularly in the premier French cooperage houses in Bordeaux, Burgundy and Cognac – trained to make superbly crafted and highly specialized barrels of white oak, unique for each wine variety and winery style. Similar satellite cooperages, and their Master Coopers, have been established in South Africa, Australia and Chile.

What separates these Master Coopers from the other craftsmen in the cooperages is their passion for the product – their desire to elicit from the oak the properties that allow it to be perfectly formed into an exquisite container for caring and nurturing the wine. You can see it in their eyes as they hold a piece of white oak: how, with a quick, knowing glance, they can discern whether it has the qualities required for a premium wine barrel and, if so, how to shape it to fit into the barrel. And one can hear it in the studied excitement of their voices as they discuss how to select the finest woods from the various forests or which oak-toasting regime will best enhance a particular wine. As Renaissance men, they can walk into a hardwood forest and talk with the loggers as they purchase the 150-year-old trees; joke with their workers in the cooperage and yet demand the extraordinary skill to fabricate barrel after barrel to the highest standards; and sit in the elegant tasting room of a winery or chateau and knowledge-ably taste and discuss the wines from their barrels with winery owners and winemakers. In short, they combine the traits of a craftsman and an artist with those of a cooperage manager and a leader.

In the various European, American and Australian cooperages, there are a number of men and women who have these qualities: people who have brought respect and prestige to their various cooper-age houses for total commitment to product quality. These select people have helped more than one winery client achieve cult status. I relate the story of one of them here. This is not to set him above the many other skilled coopers, but to present his story as an example of the paths taken to achieve such success.

Alain Fouquet is a third-generation cooper. Although Alain played in the wood shavings of his father's shop at an early age, he began his formal training at thirteen with a five-year apprenticeship at another small cooperage in Cognac, France. There, he and several other students lived and worked with a Master Cooper who became their second father. His training started with the menial work: stack-ing staves, sweeping floors and gaining the strength to lift the heavy hammers used to pound down the metal hoops that tighten the barrel. While some machinery was available, these apprentices were required to make several barrels using only hand tools. It was from this fabrication process, with both its physical and mental training, that they learned everything from the relationships between the angles and bevels on the staves to the ultimate size and capacity of the barrel. It was also during such fabrication that they could hit their thumbs and incur the small cuts and bruises one does when working – injuries which were minor and would be used to learn efficient woodworking techniques before taking on the industrial-strength machines of modern barrel production.

Alain and his fellow apprentices processed white-oak staves by hand. They would chop a slight taper from the middle of the stave to each end. Using a plane, they would then join the edges smooth, adding a critical angle – the wider the stave, the greater the angle – to fit each stave into the full circle of the barrel. The stave plane is a stout board or piece of cast iron 1.8–2.5 metres long. Alain would pass the edge of the stave down and across this blade to smooth that surface and to create the correct angle. The skill was to press down with sufficient force to plane the tough oak while holding the stave in the exact position to cut the desired angle. To obtain the perfect angle, the apprentices would check their efforts against templates for the size of barrel that they were producing. But the ultimate achievement for these apprentices, usually not attained until after many years of practice, was to be able to plane every stave angle solely by eye. It took exceptional perseverance and dedication by Alain and the other budding trainees to plane, by hand, the exact angle into the many staves required for each barrel. And then, just as importantly, to do the same for the next barrel and make it exactly the same size.

In the early 1960s, with his apprenticeship completed and a brief stint in the French army, Alain returned to another, and more modern, Cognac cooperage. There he quickly advanced to manager, due largely to his congenial personality, his barrel-crafting skills and his larger vision for the burgeoning wine-cooperage industry. While the production was focused on cognac barrels, Alain saw the growing demand for wine barrels and pushed production and sales in that direction. With his thorough knowledge of the complete barrel-crafting process, his worldly manner and his ability to speak several languages, he became the point man to expand sales world-wide. A swarthy, handsome face and a confident personality did not hinder his success. But it was the increasing demand by discerning

California winemakers for quality French oak barrels that led him to establish a second facility for the Sequin Moreau Cooperage in the Napa Valley.

In America, as in other countries, the main opportunity for a cooperage to present their barrels to the broad winemaker market takes place at an annual trade show. The most important show in the United States is held in conjunction with the meeting of the American Society of Enology and Viticulture. It was at one of these shows in the early 1980s that I first met Alain. His Sequin Moreau Cooperage was always presented in a large, well-appointed booth – plush carpets, leather chairs and salespeople in suits. At that time, most of the other coopers did well if they showed up in clean clothes. He became a role model for success in barrel sales, as his barrels were used in the finest wineries in the world. I was part of a struggling barrel-service company, but he was always helpful if we needed something, and we occasionally provided barrel repair for his company.

Alain's name was synonymous with the Sequin Moreau Cooperage, and under his leadership that brand enjoyed one of the highest reputations in the world of wine. But success led to changes in cooperage ownership. Several years ago, he quit and started his own brand of cooperage. It was an opportunity to get his hands back on the wood and have more control over the barrel craftsmanship. He is now back enjoying visiting wineries and talking with winemakers to market his own premium barrel.

A modern bronze medallion commemorating Peter Coates, the Master Cooper for T&R Theakston Ltd, Masham, North Yorkshire.

A modern cooperage, like this one in Russia, utilizes numerous machines; nonetheless, the eye of the cooper is still important.

The MINI Cooper is certainly recognized, and perhaps most might know that James Fenimore Cooper was at least an author, but sadly fewer understand the craft of a barrel cooper, or that this common English surname is derived from that ancient trade. This lack of knowledge about coopers occurs not only in the English-speaking world; just recently I had to explain to a young Frenchman that a *tonnelier* is a cooper and that barrels are built in a *tonnellerie!*

# TWELVE

# Other Barrels: Spirits, Fortified Wines and Beer

I call the barrel the heart of making whiskey.
Without the barrel, we would not have what we call bourbon.

Lincoln Henderson, Brown-Forman's whiskey-maturation expert,
*Lexington Herald*, 1999

## Whiskey Barrels

In front of me, the pyramid of used whiskey barrels was enormous. The fifteen layers of barrels created a wall at least 20 metres high, and the entire stack covered an area the size of a soccer pitch. There were several thousand barrels sitting out in that open Kentucky field awaiting disposal, with their owner anticipating sales to distillers in Scotland, Ireland, the West Indies, India or other places around the globe for the ageing of Scotch, rum, whiskey and even Tabasco sauce. If unsold, they would be sawn in half for garden planters.

My first impression, after getting over the sheer size of the pile and the numbers of barrels, was that they were just lying in the dirt, some with an open bunghole due to a missing bung. There was nothing to stop animals or insects from using them as a new home. Comparing these barrels to those used in the wine industry, where sanitation is paramount, I could not believe that these barrels could be left out in a field and then resold to age spirits for consumption.

Subsequently, I learned that sanitation was of little concern. If animals did venture near the barrels, they would leave either mighty wobbly on their legs or with a huge hangover. High-proof bourbon or whiskey thoroughly soaked the wood, and the highly alcoholic residue provided more than sufficient protection against any bacterial decay or insect invasion. A mouse or insect that did get into a barrel would be totally pickled.

After this first encounter with a mountain of whiskey barrels, my experience as the General Manager of a bourbon-barrel cooperage that had converted to a wine-barrel cooperage, plus subsequent visits

A huge stack of used whiskey barrels awaiting a buyer.

to other Midwestern and overseas whiskey-barrel cooperages and distilleries, all provided further enlightenment regarding the unique world of whiskey barrels and their uses.

## Wooden Barrels for Distillers

When confronted with the problem of how to store enough spirits to sell to others, early Greek and Roman commercial distillers situated around the Mediterranean Sea, as well as the Celtic tribes in northern Europe, realized that they needed relatively large containers. By the first millennium AD, the two obvious choices were clay vessels or wooden barrels. As with putting wine in barrels, especially those made of oak, the distillers observed that flavourful changes occurred when ageing spirits in oak barrels. Thus, over the years, wooden barrels were increasingly employed for distilled spirits.

For the early residents in northern Europe, and later in New England, no doubt a few cold winters were enough to convince them that they needed some libation to warm themselves up. Small distilleries, and their associated cooperages, sprang up in the developing cities to meet the demand. For the most part, the barrels which those cooperages produced were just the silent partner to the high-alcohol products – libations which have a mixed history.

One such product was rum, and it was in the late 1700s that the wooden barrels used for this spirit played a critical role in one of the more appalling sagas of American and British history – the slave trade. Cooperages to make barrels for distilled products were established near the emerging eighteenth-century distillers in Massachusetts and Rhode Island. The distilleries were importing raw sugar and molasses from the southern states and the West Indies and turning it into rum, which was then stored and shipped within those barrels.[1] Some of the rum would then be traded to African countries for slaves or to Britain for commercial goods which were then traded for African slaves. The slaves, packed horribly tightly in the ships' holds, were then taken to the southern United States and to the West Indies to work the cane fields. The circuit was closed by the ships returning northward with more sugar, completing the profitable, but inhumane, cross-Atlantic journeys known as the 'triangular trade'.

Later, as the early American settlers moved west, the rich virgin soils produced bountiful yields of corn, wheat and rye. The distillers, capitalizing upon this surplus, expanded to process these grains, and additional cooperages were established in the heart of the Midwest, near to the unexploited oak forests which ultimately supplied the millions of barrels these distillers required. The distillers subsequently built enormous warehouses to hold the barrels while the whiskey aged.

Whiskey barrels are used only for the ageing process. A good portion of these start their life ageing bourbon and are typically stored for four or more years in enormous bourbon warehouses. These are often eight to ten storeys high, with racks holding three tiers of barrels per floor. Due to the highly flammable nature of bourbon, especially risky in older, wood-framed warehouses, the warehouses were purposely widely separated, with the barrels containing each batch dispersed throughout a number of warehouses. The necessity of this separation became extremely clear after observing the results of the 1996 fire at the Heaven Hill Distillery in Bardstown, Kentucky, a town about 50 km south of Louisville and a centre for several bourbon distilleries. The spectacular fire, which unfortunately destroyed the distillery, the bottling plant and six of their twenty warehouses, sent a flood of burning bourbon down into a nearby creek as thousands of barrels burned. Smoke from the fire could be seen miles away. Until they rebuilt, Heaven Hill had to contract out the distilling and bottling to another company. However, most importantly, they could continue to offer the appropriately aged products, albeit in lesser amounts, due only to the fact that not all of any one batch had been destroyed.

By the late nineteenth century, cooperages in Chicago, Cincinnati, Louisville, St Louis and Milwaukee were pumping out the enormous numbers of barrels used by the beer, whiskey and other industries. Their cooperage counterparts in Great Britain and northern Europe were doing the same. Records from 1910 indicate that in the United States, approximately 91 million slack and tight barrels and kegs were produced,[2] while one million were produced in England during the same year.[3]

## Bourbon Whiskey

Bourbon is a special type of whiskey; its base is primarily corn mash rather than rye (Scotch), barley (malt whiskey) or wheat (wheat whiskey), and it is aged in new white-oak barrels for a minimum of two years. When the bourbon is removed, the used barrels are sold for ageing other products, including generic whiskies and Scotch as well as rum, ports and tequila. On emptying the barrels, if the bourbon distilleries have no immediate purchasers for their 'experienced' barrels, they stack the barrels, in literal mountains, out in fields to await a buyer – as I saw for myself in Kentucky.

Some of the barrels I observed out in that field may have come from one of the Brown-Forman Company distilleries. Brown-Forman is an established Louisville, Kentucky company, making well-known brands such as Jack Daniels, Early Times and Canadian Mist. In a discussion in 1999 of the role that barrels play in making bourbon, Lincoln Henderson, then Brown-Forman's whiskey-maturation expert, stated: 'I call the barrel the heart of making whiskey. Without the barrel, we would not have what we call bourbon.'[4]

When the bourbon barrels are being filled, the newly distilled liquid appears clear. However, if there are leaks in a whiskey barrel, the liquid, with its high sugar content, will usually seal the opening as it evaporates. Because high-proof whiskey and bourbon have a higher viscosity than wine, whiskey barrels can be made with staves that have some defects. This enables whiskey-barrel cooperages to utilize a far greater percentage of the raw staves, thus keeping the cost of the barrel much lower: around $125 per barrel versus some $300 for the least expensive, American-oak wine barrels.

The words '45 seconds of hell' formed the tag line on a 1986 advertisement for Old Grand-Dad bourbon. It referred to the 'toasting' process for whiskey barrels: heating the interior of the barrels to the point where they are actually ignited, charring the inside. Prior to this toasting, the barrels are placed on a chain conveyor and passed through a steam chamber; the steam provides the heat in a manner similar to that in which the oak scraps burn in the bending fires for wine barrels, which facilitates curving the staves. Once bent, and with truss hoops holding the staves in place, the barrels are then moved over gas fires. The barrel's interior becomes a roaring inferno (45 seconds for Old Grand-Dad barrels), requiring a quick water bath to douse the combustion. The resulting char can be between 5 and 10 millimetres deep, depending on what is requested by the distillery. The purpose of the char is to colour the bourbon, rendering its well-known amber hue, and it helps to filter the liquid. Certain distilleries, notably Jack Daniels, take another step to filter their raw bourbon by passing it through a bed of charcoal, in their case made from burning ricks of maple wood, prior to placing it in the barrels.[5]

In 1964, an interesting combination of bourbon marketing and 'good-ol'-boy' politics modified the regulations which defined 'bourbon'. By an act of the United States Congress, to name a spirit bourbon it had to be made with at least 51 per cent corn, distilled to no more than 80 per cent alcohol by volume, and was required to be aged in *new*, charred, white-oak barrels.[6] What a boon to the cooperages! The stave-mill producers and bourbon-cooperage owners were certainly smiling while enjoying their shot of bourbon over the following years. But the golden goose did not lay forever; within just twenty years the cooperages started to feel the pinch. Not only did they experience a decline in the demand for bourbon barrels, but also in the price they could charge. In 1993, one bourbon-barrel cooperage wanted to raise the price by just 50 cents per barrel. Their major distiller customer, which purchased some 200,000 barrels per year, had other ideas: the price would stay at $105 – not much more than they cost in the 1950s. The cooperage had to absorb any losses. Just as wineries are turning to oak alternatives and micro-oxygenation methods to produce less expensive wine, distillers, who still want to use barrels, also want them at giveaway prices in order to keep their production costs as low as possible.

The twentieth century saw a slow decline in the number of bourbon barrels, a significant part of the overall downturn in cooperage noted in the graph in chapter Two. In the early part of the century, around 2 million whiskey barrels were made each year.[7] Prohibition put a damper on that for several years, but with the rescinding of prohibition, production returned and peaked to over 4 million per year in 1936 and '37. By the year 2000, however, distillers required only around 1 million barrels per year, and the number of cooperages has been winnowed down from about ten to five.

All these numbers are just that – quantity. And this is another significant distinction between wine barrels and bourbon barrels. Wine cooperages seek to make better and better barrels, while bourbon cooperages strive to make them less expensively. However, to be fair, each is merely responding to their respective customers' demands. By the 1930s and '40s bourbon cooperages were using dry kilns to speed up the drying process from years to months. Steam bending and gas firing have automated production, and the man raising the barrels does so in seconds, setting up hundreds per day without ever inspecting one stave. The necessity of a team of coopers applying repairs at the end of the production line is indicative of the mechanization adding speed rather than quality.

A unique aspect of whiskey barrels is that there are actually some design specifications. By the early 1900s, with wooden barrels transporting flammable liquids such as whiskey and gasoline and toxic chemicals such as chromic and oxalic acid, it was clear that serious accidents were occurring when barrels were mishandled. In response, the United States Interstate Commerce Commission developed a set

Barrels ageing bourbon in the Buffalo Trace Distillery warehouse, Frankfort, Kentucky. Note the short chime, typical of whiskey barrels.

of rules to standardize the tight and slack barrels used for volatile and toxic products.[8] Part of the regulations stipulated that the end of the stave was to be no more than 1⅛ inch (28.6 millimetres) from the croze centre. This resulted in an unusually short chime compared to the chime on a wine barrel – one not easily broken if the barrel bounced off the truck – and overall a relatively tough, compact barrel.

This short chime constitutes a big visual difference between whiskey/bourbon barrels and wine barrels. Besides durability, with the infrequent handling of bourbon barrels there is little need for a long chime. By comparison, wine barrels are moved frequently, often monthly, for topping, cleaning and refilling, and need a longer chime to assist the labourers as they do all this shifting and lifting.

## Barrels for Other Spirits

A common factor amongst barrels used to age bourbons, whiskies, Scotch, ports, sherries and rum is that the oak is primarily American oak. And, of those, bourbon, port and sherry use new American-oak barrels, whereas the whiskies, Scotch and rum rely largely upon annual shipments of used bourbon barrels.

The second-hand bourbon barrels offer two advantages: low cost and availability. With the thousands of barrels in use annually by the spirits industry, if they approached anywhere near the current costs of wine barrels, few distillers could afford to age their products in barrels. American oak is relatively plentiful, and continued advances in the mechanization of bourbon-barrel manufacturing have kept the prices low. Port and sherry producers can use new barrels because those barrels are larger, for a cost saving per litre, plus the wages for Spanish and Portuguese coopers are relatively modest. Plus, sherry and port barrels can be used for long periods, whereas the requirement of placing 'bourbon' into new barrels dictates annual replacements. Finally, in the case of Irish whiskey and Scotch, as production of these has increased over recent decades they outstripped the domestic barrel-timber supply. So, as the product is slightly more delicate than bourbon, turning to the once-used barrel created a natural synergy.

After years of resting quietly on a rack or pallet within a bourbon warehouse, when finally emptied, the bourbon barrels are suddenly sent packing, often to the far ends of the world. Buyers from distilleries in Ireland, Scotland, India, Mexico, Canada, Taiwan, Thailand, China and the Caribbean circle like wolves, trying to get the cheapest price for the best barrels – that is, the ones with the fewest leaks and defects. The barrels are shipped in containers to distilleries in these countries to age their products, with those that can be shipped without repair termed 'select'. Those needing repair – 'seconds' which require a stave or hoop replaced or a leak spiled – are mended at local Midwestern cooperages or sometimes shipped to cooperages in Scotland or other countries for this work. Scottish coopers also enlarge some of the 53-gallon (200-litre) barrels into hogsheads – 90-gallon (340-litre) barrels – by adding extra staves and two new, larger heads. Other barrels are knocked down, with sound pieces sifted out and sold in bulk as pallets of staves, heads and hoops, which coopers in other countries put back together. The less expensive labour, combined with the cost savings in shipping several hundred potential barrels in a container instead of just 150 whole barrels, can justify this method. The used bourbon barrels which are deemed beyond repair become planters to be sold at lumber stores and garden shops.

The barrels for Irish whiskey and Scotch are primarily used bourbon barrels. However, some Scottish distilleries also include used rum barrels (which normally are also used bourbon barrels) and used sherry butts to add discreet flavour nuances to their products. Barrels for Scotch, like those for bourbon, are stored in large warehouses, often for many years. Some newer warehouses store the barrels upright on pallets, which allows for rapid access, particularly for the blended whiskies that do not benefit from extended ageing.

Whiskey is also made in Japan; however, the distilleries there primarily use barrels of American oak, made by Japanese cooperages.

The rum industry is also a purchaser of used bourbon barrels. But with rum ageing taking only six months to perhaps at most five years, the annual quantity needed is relatively small. The barrels' role in flavour enhancement for rum is slight; they are more important for the ageing component – the slow interaction with minute quantities of oxygen.

As for sherry, part of the product mix of the Kentucky cooperage where I served as General Manager included drying, processing and shipping new American-oak staves and heading to overseas cooperages. Staves and heading were shipped to various sherry coopers in Jerez, to be made into sherry butts. Spanish coopers make a range of sizes for sherry producers. Sherry is normally made using the *solera* method, where selections of wines are continuously passed through a group of barrels to age. Unlike table wine, the higher alcohol content of sherry allows the barrels to be used for many years.[9] This constant reuse of barrels minimizes their need for a continual supply of new barrels, at least relative to the whiskey industries.

Port, like sherry, starts with new *pipes*, being made largely by coopers in and around Oporto for the port-ageing cellars in Vila Nova de Gaia, the site noted at the beginning of this book. These are generally made of American oak and can also range in size, commonly from 200 to 500 litres.

Finally, Cognac and Armagnac also utilize new oak barrels, but coopered with French oak rather than American oak. The typical Cognac barrel is 300 litres, while those for Armagnac are of 500-l capacity. Most of the Cognac barrels are made of Limousin oak, from forests in the centre of France. Armagnac barrels are also made from Limousin,

A sherry barrel, with a Plexiglas head to show
the flor yeast floating atop the sherry.

as well as from oaks sourced in the foothill forests on the north slopes
of the Pyrenees. The Limousin wood has a lower tannin and higher
vanilla content compared to the more southern oaks.[10] Those southern oaks, with more tannin, add a bit of bite, which seems to be
reflected in the nutty character of Armagnac.

## Oak Ageing for Spirits and Fortified Wines

For spirits, ports and sherries, barrels play a similar ageing role to
the one they perform for wine: they soften and mellow the liquid and
infuse it with some oak characteristics. However, a big difference is
the alcohol content. Wines, with their typically lower alcohol levels
(10–14 per cent), are more susceptible to problems associated with
oxygenation – particularly spoilage due to the vinegar bacteria. The
other liquids, with their higher alcohol levels, are relatively resistant
to bacterial spoilage.

From the barrel-use point of view, too much oxygen can be
perilous for most wines, as mentioned previously. They need to be
topped up frequently, and the barrel itself is best kept horizontal to

minimize the amount of exposed surface area as the water in the wine evaporates through the wood. Leaks can allow too much oxygen to enter the barrel – thus the demand by the wineries that the barrels be absolutely sound. Uniquely, sherries utilize a flor yeast in their *solera* process. Interestingly, the wine critics Hugh Johnson and Jancis Robinson note that the flor yeast 'may be vulnerable to climate change'.[11] This yeast sits on the surface of the liquid to develop sherries' 'strong individuality',[12] and the *solera* barrels are left with a bit of 'head space'. For other higher-alcohol liquids, head space is not an issue; whiskies, and spirits in general, lose a significant amount of liquid to evaporation – often called the angels' share – during the many years of ageing. By the time the barrels are emptied, there is often only half the original content.

As I write this book, a film currently playing in town has a couple of misfit Scots attempting to steal some highly valued Scotch and blame the loss on the angels' share, which is also the name of the movie (dir. Ken Loach). That share is actually quite significant. Andrew Elder notes that some 20 million gallons of Scotch is evaporated annually as the liquid rests in the barrels,[13] and Johnson suggests that as much Cognac is lost to evaporation each year as is drunk by all of France.[14] The loss by evaporation of bourbon is similar, attested by viewing a typical bourbon warehouse, blackened by the fungus which lives on the fumes.

Finally, besides ageing, what are these spirits and the fortified and sweeter wines seeking from the oak? With the use of 'experienced' oak barrels, and of the larger-sized barrels, not much actually, as the oak flavours here are at an absolute minimum. The flavours which shine through come from the grapes themselves, the smoky peat, the malted corn and the rye, with the oak providing only a subtle undercurrent. Where new barrels are used, in bourbon and Cognac, certainly oak is a more prominent characteristic. Bourbon picks up charcoal and caramel flavours from the heavier charring on the inside of the barrel. Vanilla is also an apparent characteristic. Dill is sometimes a flavour that is mentioned, as it is associated with bourbon from the use of new American-oak barrels. However, I believe that this negative oak flavour gets highlighted only when the oak is improperly seasoned and inadequately toasted, and allowed to overwhelm a wine.

## Beer Barrels

Hundreds of years ago, beer was another obvious choice to be made and shipped in wooden barrels. And, with the exception of the few barrels built for stunts to float and bounce people over waterfalls, beer barrels were some of the toughest containers made. Due to their frequent use in transporting beer from brewery to retail outlet and back to the brewery for refilling, and the occasional pressure of the beer caused by carbon dioxide ($CO_2$), their design evolved into a short, squat keg varying from 9 to 18 gallons (34–68 litres), with the stave wood almost twice as thick as that in use for wine or whiskey barrels. They have stout hoops and are sometimes lined with pitch to resist the leaking that can arise with the extra pressure.

The boutique micro-breweries, which are proliferating around the world, have renewed the interest in beer barrels. In addition to their range of tasty ales, pilsners, IPAs (India Pale Ales), lagers, stouts, bitters and porters, some of their special brews are aged in oak barrels. Besides visiting these pubs for a pint, one can also purchase a larger quantity to take home. Buying beer in bulk has a long tradition, but usually it was not purchased in a barrel. A hundred years ago, a galvanized pail was often the standard container, filled from a wooden barrel in the local pub. Today, it is typically a plastic 1.5- or 2-litre bottle, known in New Zealand as a rigger and in other countries as a growler – again, not filled from a barrel, but from stainless-steel kegs or from tanks behind the brewery pub counter.[15]

The Celts of northern Europe were good beer consumers. With the discovery of several Roman-era breweries in Germany, it seems that they were most likely the first to use wooden barrels for their beer on a regular basis[16] – a detail which partly explains why the Germans, Dutch and Belgians have such a long history with beer. By the fourteenth century in Europe, the beer distillers and the beer-barrel coopers were institutions. As with many close business connections, down the years the brewery cooperage guilds and the breweries, which often had their own cooperages, had plenty of rough-and-tough political, and sometimes physical, battles as they fought each other over barrel prices, markets and sales.[17]

Like the oil barrel, beer barrels became labelled with distinctive sizes and names, such as the firkin for a 9-gallon (34-litre) size, kilderkin (18 gallons/68 litres) and hogshead (54 gallons/204 litres). (The 'hogshead' name is, confusingly, used for several barrel sizes and commodities.) And, also like the oil barrel, wooden beer barrels were gradually replaced by kegs of coated steel, then stainless steel, and finally aluminium containers, as their advantages over the difficult-to-clean and heavier wooden barrels became apparent.

Early European and American breweries utilized large wooden tanks and casks for mixing the beer ingredients and fermenting the result. After the fermentation, and perhaps a bit of ageing, the beer was placed in the smaller, stouter barrels. Many brews were flat, that is, without carbonation, using higher alcohol levels and hops for preservation. Others had sugar added to create a secondary fermentation which provided the $CO_2$ necessary to preserve the beer while transporting it to pubs, hotels and other establishments, where it would generally be served at room temperature.[18]

In the nineteenth century, as brewmasters and brewery owners began to learn about the need for sanitation to ward off the disadvantages brought by microbacterial contamination, they sought out measures to clean the barrels. A metal bung fitting was devised,

Note the thick staves and wide, strong steel hoops of these beer barrels. The tap would be placed in the small hole in the head of the barrel.

both to tap the beer and allow for sanitizing the barrel after use. Despite barrels being lined with various substances, the overall advantages of metal containers meant that they overtook wooden kegs in the late twentieth century.

However, since the 1960s or so, there has been a renewed interest from micro-breweries in utilizing wooden containers for ageing beer. Used whiskey or wine barrels, because they are widely available and the least expensive, are the common choice.[19] While these used barrels have little tannin left, according to brewmaster Garrett Oliver, they can provide some vanilla and caramel flavours.[20] Barrels used for Scotch can provide a bit of peat flavour, and some of the grain flavours from bourbon may be apparent, especially if there is a bit of bourbon left in the barrel. Wine barrels used for red wines can turn the beer a rosy colour, while aromatic wines such as Riesling can express some fruit and berry flavours in the resulting beer (although I believe that finding barrels previously used for Riesling would be extremely difficult because it is rarely aged in barrels).[21]

Wood chips are also used in the brewing industry. Beechwood chips, for example, are used by one of America's largest breweries, Budweiser. Despite their claim of 'beechwood ageing',[22] the chips are not used to age or flavour their beer; prior to use, the chips are steamed to remove any resins and oils. Instead, they are used for their relatively large surface areas, which become perfect attachment sites for brewing yeast during the fermentation process occurring within the stainless-steel tanks. From the relatively permanent bases on the beech chips, the yeast can process some of the more astringent compounds, thereby mellowing the beer.[23] Previously, the relatively rough interior of a wooden barrel or wooden tank would have provided this same supporting structure.

Beer barrels seem to be associated with more colloquial terms and phrases than other types of barrels. The phrase, too well-known to college students, 'to tap the keg' has its origins in beer kegs. Early beers had a low carbon-dioxide content, and only having slight pressure enabled a wooden bung to be used.[24] In the 1950s, a metal bung was patented to 'tap' the beer kegs. This device proved useful for dispensing beer in the pub and restaurant, and at college and – a U.S. institution – tailgate parties. Initially, the tap included a pump to pressurize the beer. Later, carbon dioxide was added to the beer itself to provide the pressure. And perhaps due to the typically uncomplicated conviviality associated with beer drinking, as opposed to that with other alcoholic drinks, a number of songs and sayings have been created, several related directly to the beer barrels. The words for 'Roll out the barrel, and we'll have a barrel of fun', were incorporated into the 'Beer Barrel Polka', an extremely popular song after the Second World War.

Other sayings, related to barrels but not necessarily to beer, include 'over a barrel', 'money on the barrel head', 'barrel of laughs' and 'scraping the bottom of the barrel'. The first two resulted from personal interactions while trading materials or supplies; the third might be a variation on 'a barrel of monkeys'; and the last came from searching for the last bit of money in barrels when they were used to ship coins.[25]

# THIRTEEN

# Oak Flavouring: Oak Alternatives and Barrel Shaving

Oak alternative products are like any tool. The more control that tool allows a careful artist, the better that artist will be.

Dr Richard Carey, 'Use of Oak Alternatives in Modern Winemaking', 2009

## Oak Alternatives

In 2001, a wine selling for the incredibly low price of $2 per bottle hit the shelves of the American boutique grocery chain Trader Joe's. The brand was Charles Shaw, but it quickly gained the nickname 'Two Buck Chuck'.[1] Besides the remarkable price, it actually tasted as though it had been aged in oak barrels. The story of how a wine could sell so cheaply – about the controversial Fred Franzia and his Bronco Wine Company, which made the wine – is beyond the scope of this book. But we can look at how the wine was oaked.

Bronco is one of California's huge wineries, processing enormous quantities of wine under a number of labels at its plant in the Central Valley. Most of it, including Two Buck Chuck, is made in stainless-steel tanks measured not in the thousands of litres but in the hundreds of thousands of litres. For the Chuck wine, the use of oak alternatives, in this case mostly toasted oak planks placed in the tanks, lends the Chardonnay and Cabernet a hint of oak flavour. The planks are the length and width of staves, but only about 6 millimetres thick. Mimicking the interior toasting of a barrel, they have been toasted by oak scrap fires or by infrared heaters. The planks can be placed in a tank loose, but a system to contain them – keeping them from floating to the top and out of the way of manways/access ways and drain fittings – works best.

When I visited the Bronco winery, their winemakers were using a routine of two batches of wine, each in the tank for one month, per load of oak planks. For ageing the wine, some was actually placed into lower-cost American-oak barrels for a short period. The bulk

Some of the shapes and sizes of oak alternatives available. Clockwise from left: small toasted chips, powder, and toasted and untoasted large chips.

was aged in tanks utilizing microOx. Bronco was able to produce this absurdly inexpensive wine by processing enormous volumes. To supplement their own huge quantity of low-cost Central Valley grapes, they purchased discounted wine on the bulk wine market. And by utilizing these alternative winemaking techniques with minimal handling, they were able to keep the costs amazingly low.

The proliferation of oak alternatives is one area which has seen enormous growth as wineries struggle to control costs while still providing wines that taste like they have been in barrels. The oak alternatives are available in a huge range of shapes (powder, chips, balls, cubes, planks and so on), but all are simply pieces of oak which have been dried and toasted in a manner similar to the wood of a barrel. They are placed in the wine, usually while it is resting in stainless-steel tanks as outlined above, to provide oak flavour components without the cost of actually ageing the wine in barrels. This can be done in conjunction with micro-oxygenation, for the ageing aspect, and/or in some actual barrels using new, used and/or shaved barrels (more on shaving below).

What started with using just raw oak chips, placed in mesh bags to become a 'tea bag' within a tank, or sprinkled on the grapes in the crusher, has evolved into a mini-industry within the larger cooperage world. Most wine-cooperage companies now offer an array of proprietary chips, blocks, cubes, wands and slats to meet any requirement, or fantasy, of the winemaker. On product labels, wine-marketing gurus are now using phrases such as 'oak maturation' or 'oak influence', to suggest the deployment of oak barrels without actually mentioning the use of oak alternatives.[2]

Using oak chips to flavour wine may not be new, but it was certainly ramped up in the 1980s and '90s, when some of the huge Central Valley wineries starting ordering millions of kilograms of chips per year.[3] California's winemaking history and, by extension, its traditions and regulations, are but a few years old when compared to Europe's two millennia plus of experience; and adding oak flavouring in alternative ways is not prohibited. As the wineries proliferated, and competition for consumers and shelf space became intense, new wine styles, and especially less expensive methods to achieve the traditional styles, became commonplace. This certainly put France in a difficult position; there, it was unlawful to use chips in the winemaking process, while American wines could be made with the taste and feel of European barrel-aged wines at a fraction of the cost. Placed side by side, the French wines may have tasted better, but not all consumers experienced their wines that way. It is only recently that France has modified its stance on the use of chips and oak alternatives in order to level the playing field.[4]

The first chips were basically raw oak, but as their use grew, winemakers started requesting toasted chips. Later, as they gained experience using the smaller pieces, they asked for larger individual

Installing a barrel insert: a set of toasted slats attached to food-grade plastic rings which hold them in place.

chunks of wood in order to slow down the extraction process and improve the flavour integration. Thus each cooperage developed their own proprietary range of shapes, and most offered all shapes in light, medium and heavy toast levels.

Besides sticking planks of toasted oak and 'tea bags' of chips into tanks, whole sets of toasted slats, termed 'inserts', are available to stick into barrels. This is an alternative to shaving the inside of the barrel: a rejuvenation of the wood by inserting enough slats to equal the interior surface area of the original barrel.

Using oak alternatives is becoming a standard procedure in many wineries, although there are no reliable statistics to document this trend, nor is it readily publicized by the wineries. However, this should not necessarily be considered bad practice. Throughout the ages, winemaking techniques have constantly evolved, mostly for the better, as viticulture and oenology practices improved and consumer preferences changed. The use of oak alternatives is really not very different from any other shifts in winemaking which have occurred throughout the aeons: using large tanks or small barrels, using one type of oak as opposed to another, fermenting in barrels or in tanks, using natural or prepared yeasts or adjusting the acids or sugars of a wine. Obviously each produces a different style, but it is ultimately up to us, the consumers, to indicate which we prefer.

And apparently many wine drinkers prefer the ways in which wine is currently being made. Globally, wine sales have increased 3.4 per cent from 2007 to 2011 (driven largely by consumers in the United States, China and Russia).[5] Much of the wine, especially the reds, is made with oak alternatives and is destined for supermarket shelves as opposed to fine-wine shops. And, due to convenience, lack of storage space and cost, more and more consumers are purchasing their wine in supermarkets, or from unlikely places such as Starbucks, or online from numerous retailers.[6]

## Using Shaved Barrels

With the increasing use of barrels worldwide, many wineries, both large and small, cannot afford the significant amounts of money needed for new barrels as a way to oak their wines. While both scraping the inside of used barrels to expose new oak, now termed 'shaving', and adding oak-wood chips of various sizes and shapes have proliferated in the past 50 years, it is probable that they have been utilized in some form or another for the past 2,000; just without as much fanfare.

The first shaving of barrels was literally scraping by hand with an adze or hatchet. Later, chains were utilized – basically to beat the inside of the barrel and knock off the accumulated tartrates. As mechanization was introduced into cooperages, an alternative system was developed to dismantle the barrel, run each stave through a curved-blade planer and then reassemble the barrel. The planer allowed for the operator to bypass the ends of the staves, thereby preserving the croze and chime sections and planing only the surface of the middle of the stave. The heads could be planed utilizing a standard, straight-blade planer. This process removes about 5 millimetres, exposing relatively new oak. Before the heads were replaced, the barrel might also have been placed on a fire to simulate the toasting it received when originally built.

At least by the 1970s, machines were developed to shave the inside of the barrel by removing the heads but leaving the body of the barrel intact. An experienced three-man team could shave 100 barrels per day. These machines consisted of a router, suspended on an arm with which it could reach inside the barrel, and wheels on which the barrel sat and rotated as the router cut an overlapping spiral through it to expose new oak.

Like the eighteenth- and nineteenth-century itinerant coopers who made a living by peddling their wooden products and repairing broken wooden buckets, pails and barrels, the barrel-shaving machine and operation is taken from winery to winery. The Napa Valley company for which I worked made frequent trips to service wineries in California, Oregon and Washington and occasionally to wineries in the eastern states as well. One time, the shaving machine was even shipped to Israel to provide the service for several wineries in the Golan Heights.

As the cost of wine barrels has risen – close to 400 per cent over the past 35 years – and more and more wineries want to extract every last vestige of oak, continued efforts have been made to develop different barrel-shaving systems. Subsequent advancements in 'barrel shaving' have incorporated different types of router blades and wire whisks, an extremely high-powered water blaster and the use of rice-sized, dry-ice pellets to pepper the interior surface explosively, blasting away spent oak.

All of these shaving processes expose oak which has not had its properties totally changed. The extra step of retoasting – by oak scrap fire or infrared heating elements – does reprovide at least a modicum of the changes to the oak which occurred when the barrel was originally built. One drawback with the retoasting is the heating of any residual wine in the barrel wood. Despite what cooperage shaving advertisements note, wine penetration extends into at least half of the stave or head board – deeper than the 5–8 millimetres normally removed by shaving processes. And while the vast majority of spent oak is removed in the shaving process, the small amount of wine remaining within the wood could, if burned in the toasting process, add some bitterness to the new wine. However, perhaps the honourable attempts at fully utilizing barrels, as opposed to selling them for planters after just a few years of use, plus the cost savings to the wineries, justify the shaving and retoasting processes.

# FOURTEEN

# Cooperage: The Bigger Picture

 What happens to barrels after they have been used *and* reused? Bourbon barrels initially get recycled for Scotch, generic whiskies and rum, while wine barrels generally become planters. And after that?

Unfortunately, as is the case for much of our technological progress, there are negative consequences. Currently, the industrial world's incredible technical skills, which produce a cornucopia of foods and products, all artfully wrapped and protected, also create an enormous amount of wasteful packaging. With so much of the packaging designated for one-time use, the downside is that the packaging creates millions of tonnes of garbage. While most goes into landfills and some gets recycled, a significant portion sadly ends up carelessly tossed out, scarring our roadways, parks, rivers, beaches, towns and cities. So much plastic has been disposed of improperly that huge gyres are now found floating in the middle of our oceans and seas, filled with thousands of bits of flip-flops, Styrofoam, plastic drink bottles and other flotsam and jetsam – all by-products of our technological 'progress'.

In past eras, barrels were also part of the improperly disposed rubbish. Their useful life over, some of the barrels were inevitably tossed out into the streets or rivers, or if at sea, thrown overboard. Sea-story author Joan Druett cites the tale of throwing a barrel overboard as target practice for the ship's cannons.[1] However, for generic barrels at least, the wood – the staves and heading – could also have been burned for fuel or heat. Otherwise, if the barrel or its wood was left on the ground, they would eventually decompose, with less harm to the environment than the current plastic bags, bottles and wrappings. And the early hoops, which were wooden, were also biodegradable and decomposed relatively quickly.

The shift into the use of iron, and later steel, hoops pushed the barrel out of the 'eco-packaging' category. Those steel hoops, whether having fallen off a discarded barrel or been irresponsibly thrown onto the ground, would at the very least have been an eyesore and nuisance or, worse, a potential danger to people or animals, as they tend to flip up if stepped on, whacking one's shin in the process. However, at least some of the steel hoops were recycled for use on other barrels, or as tools such as hoes, knives, machetes and scrapers. For just this reason, early explorers traded those hoops with indigenous peoples whose culture did not include making iron products.

Another disposal problem was created when the barrels were lined with tars and oils or their wood impregnated with petroleum products. Many petroleum containers were shipped by river barge, with those that did leak fouling the waterways. After use, many of those barrels were probably burned, further adding to the air pollution. And it is probable that some of the barrels also just broke down, lying in some alleyway or ditch, and thus contributed to the general clutter and polluted the runoff.

The graveyard for today's barrel is usually when they are cut in half and used as planters, slowly decomposing into the very soil they attempt to contain. Some artists and craftspeople do attempt to make art and furniture from the staves and hoops. Unfortunately, they eventually find that their creative efforts rarely justify the prices they need to ask for their struggles to process barrels' odd shapes, lack of standardized sizes and awkward components.

Given what we now know of the advantages and disadvantages of wooden barrels, one has to question whether it might be beneficial to use more barrels in today's world. My answer is that I am not so sure. Many plastic, paper and lined cardboard products are much more convenient than wooden barrels. Probably the better question would be: could we use more bulk containers rather than a multitude of individualized containers? Here I can definitely answer: yes!

Regarding bulk containers, there isn't much doubt that our fast-food mentality promotes the use of far more packaging than

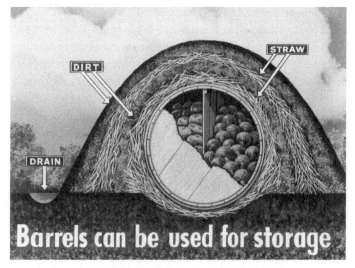

Barrels can be used for storage

Some barrels are reused in creative ways.

necessary. It has finally reached the point where the global retailer Walmart is actively encouraging its suppliers to minimize their packaging. In relating how they are attempting to reduce packaging, Walmart's website noted:

> We worked with one of our toy suppliers to help them reduce packaging on 16 items. As a result, we were able to use 230 fewer shipping containers to distribute their products, saving about 356 barrels of oil and 1,300 trees. By broadening this initiative to 255 items, we expect to save 1,000 barrels of oil, 3,800 trees and millions of dollars in transportation costs.[2]

Besides the impressive reduction in shipping containers and savings in oil and trees, this statement is interesting and appropriate for this book. Walmart used the term 'barrels' for the barrels of oil saved, coming from the days when oil was actually shipped in barrels. And would a better use of the saved trees be to make wooden barrels?

Another worthy sideline on Walmart's website was related to the challenge of reducing waste for their electrical wire suppliers.

Traditionally, electrical wire is shipped and dispensed on steel, plywood or cardboard reels. Somewhat like the barrel, in the sense of its inability to be recycled, these reels have almost no other function and end up clogging landfills with their difficult-to-thoroughly-flatten shape. Accepting this challenge, one supplier developed a cardboard wire carrier and dispenser which was totally recyclable.[3]

The sudden drop on the right-hand side of our graph depicting the wooden barrel's history – illustrating the dramatic decline in its use – reveals analogies to the decline of the amphora. As for the more current example of the shift from physical newspapers and books to digital media, a local public panel discussion in my town highlighted the rationale causing this change rather well.

The panel consisted of a large-format book publisher, a bookshop owner, a newspaper editor and a librarian. Not unexpectedly, all except the librarian were still somewhat biased towards protecting existing hardcopy technology as they struggled to find their new niche in the digital world. What seemed to be significant in this discussion were the two reasons, offered by the man from the local public library, for the shift from printed material to reading and gathering the same information in digital formats. He believed that it was a combination of both the push by technology and society's desire for change which

These wire reels, like barrels, are difficult to recycle, or to dispose of in landfill.

had fuelled this switch. While physical printed matter is still quite important, most of us are also utilizing some or all of the digital formats.

Certainly the push by technology shifted barrel use. We've examined a number of cases where containers made of the new materials – such as steel, aluminium and later plastic – rapidly became substitutes for barrels due primarily to their ability to be cleaned and to be relatively free from bacterial contamination. And consumers, finicky in their purchasing habits, have, for a variety of reasons, opted for smaller portions. Both of these general trends eroded the need for wooden barrels.

The demise of the amphora occurred within several hundred years; the significant decline in barrel use has occurred within the past century; and now we are barely a few years, or at most a little over a decade, into this change from physical books and magazines to digital reading. However, while hard copies may always be around, certainly the digital formats will continue to make inroads into our leisure reading and into how we obtain our news and information. We have seen this same trend with the demise of wooden barrels, from a multi-functional bulk container to their current use primarily for premium wines and spirits.

The shift from barrels to other materials is on the far side of the curve, whereas the swing from reading hard copies to reading digitally is just beginning. Nevertheless, with a 2,000-year history of necessity, wooden barrels will, most likely, continue to be used for some years to come – albeit, sadly, in ever-diminishing numbers.

# References

## Introduction

1 Simon S. James, *Exploring the World of the Celts* (London, 1993),
   p. 7.
2 Hugh Johnson, *The World Atlas of Wine* (New York, 1995), p. 216.

## ONE
## Need: Why Wooden Barrels?

1 Ron Lambert and Gail Henry, *Taranaki: An Illustrated History* (Auckland,
   2000), p. 123.
2 Franklin E. Coyne, *The Development of the Cooperage Industry in the United
   States, 1620–1940* (Chicago, 1940), p. 33; Lambert and Henry, *Taranaki*;
   and Jean-Marc Soyez, *Les Ebenistes du Vin*, trans. Michael Mills (Lormont,
   France, 1991), p. 36.
3 Kenneth Kilby, *The Cooper and His Trade* (Fresno, CA, 1977), p. 135.
4 Coyne, *The Development of the Cooperage Industry in the United States*, p. 11;
   and Jennifer Kennedy, 'Brief History of Cod Fishing',
   www.marinelife.about.com, accessed 15 May 2011.
5 Clyde L. MacKenzie Jr, 'History of Oystering in the United States and
   Canada, Featuring the Eight Greatest Oysters', *Marine Fisheries Review*,
   LVIII/4 (1996).
6 Diana Twede, 'The Cask Age: The Technology and History of Wooden
   Barrels', *Packaging Technology and Science*, 18 (2005), p. 254.
7 Coyne, *The Development of the Cooperage Industry in the United States*,
   p. 17.

8 Ibid., p. 33.

9 Soyez, *Les Ebenistes du Vin*, p. 10.

10 Coyne, *The Development of the Cooperage Industry in the United States*, p. 7.

## TWO
## Evolution: From Buckets to Barrels

1 Diana Twede, 'The Cask Age: The Technology and History of Wooden Barrels', *Packaging Technology and Science*, 18 (2005), p. 254.

2 Ibid.

3 Ibid.

4 Ian Morris, *Why the West Rules – For Now* (London, 2010).

5 Kenneth Kilby, *The Cooper and His Trade* (Fresno, CA, 1977), p. 172.

6 Barry Cunliffe, *The Ancient Celts* (Oxford, 1997), p. 33.

7 Morris, *Why the West Rules*, p. 130.

8 Ibid., p. 131.

9 Brian Fagan, *The Long Summer: How Climate Changed Civilization* (New York, 2004), p. 196.

10 Diana Twede, 'Commercial Amphoras: The Earliest Consumer Packages?', *Journal of Macromarketing*, 22 (2002), p. 99.

11 Ibid.

12 Morris, *Why the West Rules*, p. 131.

13 Barry Cunliffe, ed., *Prehistoric Europe: An Illustrated History* (Oxford, 1996), pp. 417–18.

14 Morris, *Why the West Rules*.

15 John Davis, *The Celts* (London, 2000), p. 68.

16 Morris, *Why the West Rules*, p. 33; and Twede, 'The Cask Age', p. 254.

17 Twede, 'Commercial Amphoras', p. 105.

18 Ibid., p. 99.

19 Ibid., p. 102.

20 Ibid., p. 103.

21 Ibid., p. 100.

22 Leslie Adkins and Roy Adkins, *Introduction to the Romans* (London, 1991), p. 102.

23 Twede, 'Commercial Amphoras', p. 98.

24 Ibid., p. 103.

25 Morris, *Why the West Rules*, p. 28.

26  Twede, 'Commercial Amphoras', p. 99.
27  Ibid.


**THREE**
## Celts: A Nexus of Skills and Technology

1   John Bostock and H. T. Riley, trans., *The Natural History of Pliny*, vol.
    III (1855), www.books.google.co.nz, accessed 1 September, p. 268.
2   Ibid.
3   Peter Berresford Ellis, *The Ancient World of the Celts* (London, 1998),
    p. 153.
4   Simon S. James, *Exploring the World of the Celts* (London, 1993), p. 114.
5   Ellis, *The Ancient World of the Celts*, p. 153.
6   Kenneth Kilby, *The Cooper and His Trade* (Fresno, CA, 1977), p. 93.
7   Barry Cunliffe, *The Ancient Celts* (Oxford, 1997), p. 57.
8   Ellis, *The Ancient World of the Celts*, p. 155.
9   Ibid., p. 9.
10  Ibid.
11  Cunliffe, *The Ancient Celts*, p. 6.
12  Garrett Oliver, ed., *The Oxford Companion to Beer* (Oxford, 2012),
    p. 388.
13  James, *Exploring the World of the Celts*, p. 56.
14  Cunliffe, *The Ancient Celts*, p. 57.
15  John Davis, *The Celts* (London, 2000), p. 18.
16  Ibid., p. 19.
17  Davis, *The Celts*.
18  Davis, *The Celts*, pp. 18–19; and Ellis, *The Ancient World of the Celts*,
    p. 137.
19  Bostock and Riley, *The Natural History of Pliny*, p. 269.
20  Donald H. Menzel, *A Field Guide to the Stars and Planets* (Boston, MA,
    1964), p. 112.
21  Cunliffe, *The Ancient Celts*, p. 57.
22  Ellis, *The Ancient World of the Celts*, p. 10.
23  Brian Fagan, *The Long Summer: How Climate Changed Civilization*
    (New York, 2004), p. 198.
24  James, *Exploring the World of the Celts*, p. 146.
25  Oliver, *The Oxford Companion to Beer*, p. 388.

26 'Three Millennia of German Brewing',
   www.germanbeerinstitute.com, accessed 21 November 2012.

27 Ian Morris, *Why the West Rules – For Now* (London, 2010), p. 232.

28 Ibid., p. 233.

29 R. F. Tylecote, *The History of Metallurgy* (London, 1976), p. 40.

30 Ibid.

31 Davis, *The Celts*, p. 46.

32 Leslie Adkins and Roy Adkins, *Introduction to the Romans* (London, 1991), p. 81.

33 Cunliffe, *The Ancient Celts*, p. 43.

34 Ellis, *The Ancient World of the Celts*, p. 129.

35 Ibid.

36 Ibid.

37 Cunliffe, *The Ancient Celts*, p. 53.

## FOUR
## Romans: Employing the Barrels for Trade

1 Norman Davies, *The Isles* (London, 1999), p. 137.

2 Umberto Eco, *Serendipities; Language & Lunacy*, trans. William Weaver (London, 1998), p. 68.

3 John Bostock and H. T. Riley, trans., *The Natural History of Pliny*, vol. III (1855), www.books.google.co.nz, accessed 1 September 2011, p. 268.

4 Stephan Dando-Collins, *Caesar's Legion: The Epic Saga of Julius Caesar's Elite Tenth Legion and the Armies of Rome* (New York, 2002), p. 13.

5 Ibid., p. 63.

6 W. A. McDevitte and W. S. Bohn, trans., *The Works of Julius Caesar* (1869), www.sacred-texts.com, accessed 3 October 2012, Chap. 8, Sect. 42.

7 McDevitte and Bohn, *The Works of Julius Caesar*, Chap. 8, Sect. 46; and Dando-Collins, *Caesar's Legion*, p. 63.

8 Jean Taransaud, *Le livre de LA TONNELLERIE* (Paris, 1976), p. 18.

9 Kenneth Kilby, *The Cooper and His Trade* (Fresno, CA, 1977), p. 96.

10 Mike Dalka, 'What Things Cost in Ancient Rome', www.constantinethegreatcoins.com, accessed 8 September 2012.

11 Alexander Canduci, *Triumph and Tragedy: The Rise and Fall of Rome's Immortal Emperors* (Millers Point, NSW, 2010), p. 61.

12  Kilby, *The Cooper and His Trade*, p. 98.

13  Dalka, 'What Things Cost'.

14  Gérard Aubin, Sandrine Lavaud and Philippe Roudié, *Bordeaux: Vignoble millénaire* (Bordeaux, 1996), p. 47.

15  Davies, *The Isles*, p. 137.

16  Canduci, *Triumph and Tragedy*, p. 37.

17  Taransaud, *Le livre de LA TONNELLERIE*, p. 18.

18  Diana Twede, 'The Cask Age: The Technology and History of Wooden Barrels', *Packaging Technology and Science*, 18 (2005), p. 254.

19  Peter Berresford Ellis, *The Ancient World of the Celts* (London, 1998), p. 153; Hugh Johnson, *The World Atlas of Wine* (New York, 1995), p. 13; and Joan Liversidge, *Britain in the Roman Empire* (Oxford, 1969), pp. 51–2.

20  Liversidge, *Britain in the Roman Empire*, pp. 51–2.

21  Barry Cunliffe, *The Ancient Celts* (Oxford, 1997), p. 150.

22  Author's visit to Pompeii exhibit in Melbourne, Australia, 4 August 2009; see also http://theinquisition.eu, accessed 15 August 2012.

23  Twede, 'The Cask Age', p. 255.

24  Ibid.

25  Charles Phillips and D. M. Jones, *The Everyday Life of the Aztec and Maya* (London, 2007), p. 97; and Jen Green, Fiona MacDonald, Philip Steele and Michael Stotter, *The Encyclopedia of the Ancient Americas* (London, 2000), pp. 97, 143.

26  R. F. Tylecote, *The History of Metallurgy* (London, 1976), p. iv.

## FIVE

## Middle Ages: A Surge in Barrel Use

1  Gérard Aubin, Sandrine Lavaud and Philippe Roudié, *Bordeaux: Vignoble millénaire* (Bordeaux, 1996), p. 9.

2  Ibid.

3  Norman Davies, *The Isles* (London, 1999), p. 137.

4  Aubin, Lavaud and Roudié, *Bordeaux*, p. 12.

5  Anne Crawford, *A History of the Vintners' Company* (London, 1977), p. 16.

6  Aubin, Lavaud and Roudié, *Bordeaux*, p. 12.

7  Ibid.

8 Ibid.

9 Kenneth Kilby, *The Cooper and His Trade* (Fresno, CA, 1977), p. 97.

10 Diana Twede, 'The Cask Age: The Technology and History of Wooden Barrels', *Packaging Technology and Science*, 18 (2005), p. 256.

11 Bob Dean, 'A 3rd Century AD Gallo-Roman Trading Vessel from Guernsey aka The St. Peter Port Wreck or "The Asterix Ship"', www.swan.ac.uk, 2004, accessed 26 September 2012.

12 Peter H. Blair, *An Introduction to Anglo-Saxon England*, 3rd edn (Cambridge, 2003), p. 283; and Kilby, *The Cooper and His Trade*, p. 99.

13 Dean, 'The Asterix Ship'.

14 Leslie Adkins and Roy Adkins, *Introduction to the Romans* (London, 1991), p. 102.

15 Dean, 'The Asterix Ship', p. 12.

16 Twede, 'The Cask Age', p. 256.

17 Jean-Paul Lacroix, *Bois de Tonnellerie: De la forêt à la vigne et au vin* (2006), http://books.google.co.nz, accessed 15 September 2012, p. 57.

18 Aubin, Lavaud and Roudié, *Bordeaux*, p. 13.

19 Jean-Marc Soyez, *Les Ebenistes du Vin*, trans. Michael Mills (Lormont, France, 1991), p. 11; and Jean Taransaud, *Le livre de LA TONNELLERIE* (Paris, 1976). Taransaud notes that the word for the axe to taper the staves is *la doloire*, while the man who actually uses the axe is *le doleur*, which has a number of variant spellings: *le doileur, le dolleur, le dolleu, le doleure* and *le doleur*, p. 66.

20 Aubin, Lavaud and Roudié, *Bordeaux*, pp. 13, 103; Lacroix, *Bois de Tonnellerie*, pp. 35–9, 52; and Taransaud, *Le livre de LA TONNELLERIE*, pp. 72–3, 76–7, 111, 193, 215.

21 Lacroix, *Bois de Tonnellerie*, p. 51.

22 Ibid.

23 Lacroix, *Bois de Tonnellerie*, p. 55.

24 Kilby, *The Cooper and His Trade*, p. 110.

25 Lacroix, *Bois de Tonnellerie*, p. 57.

26 Catherine Rachel Pitt, 'The Wine Trade in Bristol in the Fifteenth and Sixteenth Centuries', www.bris.ac.uk, 2006, accessed 20 October 2012, pp. 23–9.

27 Ibid., p. 77.

28 Crawford, *A History of the Vintners' Company*, p. 15.

29  Ibid.

30  Lacroix, *Bois de Tonnellerie*, p. 57.

31  Pitt, 'The Wine Trade in Bristol in the Fifteenth and Sixteenth Centuries', p. 22.

32  See www.newbordeaux.com, accessed 28 March 2012.

33  Lacroix, *Bois de Tonnellerie*, p. 57; and Twede, 'The Cask Age', p. 254.

34  Aubin, Lavaud and Roudié, *Bordeaux*, p. 46.

35  Ibid.

36  Pitt, 'The Wine Trade in Bristol in the Fifteenth and Sixteenth Centuries', p. 112.

37  Lacroix, *Bois de Tonnellerie*, p. 58.

38  Ibid.

39  Ibid.

40  Margery K. James, 'The Medieval Wine Dealer', *Explorations in Entrepreneurial History*, x/2 (1957), pp. 45–53.

41  Crawford, *A History of the Vintners' Company*, pp. 23–4.

42  James, 'The Medieval Wine Dealer', p. 48.

43  Ibid.

44  Crawford, *A History of the Vintners' Company*, p. 15.

45  Ibid.

46  Twede, 'The Cask Age', p. 258.

47  Ibid.

48  The Worshipful Company of Coopers, www.coopers-hall.co.uk, accessed 10 January 2012.

49  Ibid.

50  Kilby, *The Cooper and His Trade*, p. 108.

51  Ibid., p. 105.

52  Ibid., p. 106.

53  Crawford, *A History of the Vintners' Company*, p. 13.

54  Kilby, *The Cooper and His Trade*, p. 111.

55  Kilby, *The Cooper and His Trade*, p. 127; and Sylvia L. Thrupp, *The Merchant Class of Medieval London* (1988 [1948]), http://books.google.co.nz, accessed 9 October 2012, pp. 3–4.

56  Taransaud, *Le livre de LA TONNELLERIE*, p. 213.

57  Ibid.

58  Ibid.

59  Kilby, *The Cooper and His Trade*, p. 111.

SIX
## Parallels: Wooden Barrels and Wooden Boats

1  Franklin E. Coyne, *The Development of the Cooperage Industry in the United States, 1620–1940* (Chicago, 1940), p. 9.

2  Ibid.

3  William Bryant Logan, *Oak: The Frame of Civilization* (New York, 2005), p. 194.

4  John Guzzwell, *Modern Wooden Yacht Construction: Cold-molding, Joinery, Fitting Out* (London, 1979), p. 3.

5  Ibid.

6  Guzzwell, *Modern Wooden Yacht Construction*, p. 5.

7  Logan, *Oak*, p. 173.

8  Guzzwell, *Modern Wooden Yacht Construction*, p. 40.

9  Kenneth Kilby, *The Cooper and His Trade* (Fresno, CA, 1977), p. 71.

10  Andrew Lambert, 'French Battleship Napoléon (1850)', www.cityofart.net, accessed 10 March 2011

11  Logan, *Oak*, p. 205.

12  Kilby, *The Cooper and His Trade*, p. 72.

13  Maynard Bray, 'Wood for Boats', in *The Wooden Boat Series: Planking and Fastening*, ed. Peter H. Spectre (Brooklin, MA, 1996) p. 64.

14  Ibid.

15  Diana Twede, 'The Cask Age: The Technology and History of Wooden Barrels', *Packaging Technology and Science*, 18 (2005), p. 254.

16  Author's visit to the site of *Edwin Fox*, Picton, New Zealand.

17  George S. Nares, *Seamanship*, 2nd edn (Facsimile edn, Woking, 1979 [1862]), p. 79.

18  Peta Raggett, maritime author and model shipbuilder, personal interview and emails, 21 April 2010, Nelson, New Zealand.

19  Ibid.

20  Twede, 'The Cask Age', p. 255.

21  Harry Morton, *The Wind Commands: Sailors and Sailing Ships in the Pacific* (Dunedin, NZ, 1975), p. 291.

22  Twede, 'The Cask Age', p. 255.

23  Morton, *The Wind Commands*, p. 289.

24  John Gascoigne, *Captain Cook: Voyager between Worlds* (London, 2007), p. 99.

25  Joan Druett, maritime author, personal correspondence via email of 18 and 19 May 2010.

26  Gascoigne, *Captain Cook*, p. 16.

27  Nicholas Thomas, *Cook: The Extraordinary Voyages of Captain James Cook* (New York, 2003), p. 416.

28  Gascoigne, *Captain Cook*, p. 35; and Thomas, *Cook: The Extraordinary Voyages*, p. 164.

29  Ibid., page 53.

30  Ibid., page 57.

31  Thomas, *Cook: The Extraordinary Voyages*, p. 164.

32  Ibid., p. 335.

33  Gascoigne, *Captain Cook*, p. 57.

34  James Cook, *A Voyage towards the South Pole and Round the World* (1777), http://books.google.co.nz, accessed 28 April 2010, p. 94.

35  Morton, *The Wind Commands*, p. 302.

36  Ibid.

37  Morton, *The Wind Commands*, p. 289.

38  Thomas, *Cook: the Extraordinary Voyages*, p. 416.

39  Druett, personal correspondence.

40  Morton, *The Wind Commands*, p. 289.

41  John H. St J. de Crèvecoeur, *Letters from an American Farmer* (1782), http://xroad.virginia.edu, accessed 9 May 2011, pp. 158–9.

42  Granville Allen Mawer, *Ahab's Trade: The Saga of South Seas Whaling* (Leonards, NSW, 1999), p. 289.

43  Ibid.

44  Mawer, *Ahab's Trade*, p. 74.

45  William H. Chappell, entry of 23 March 1853 from personal diary while serving as ship's cooper on board the whaling ship *Saratoga*; retrieved via microfilm from the Alexander Turnbull Library, Wellington, 10 May 2011.

46  Mawer, *Ahab's Trade*, p. 9.

47  Ibid., pp. 74–5.

48  Ibid., p. 76.

49  Curator, Russell Maritime Museum, Russell, NZ, email of 17 June 2011.

50  Druett, personal correspondence.

51  Ibid.

## SEVEN
## Organizations: From Guilds to Cooperages

1 Kenneth Kilby, *The Cooper and His Trade* (Fresno, CA, 1977), p. 97.
2 Edwin S. Hunt and James M. Murray, *A History of Business in Medieval Europe, 1200–1550* (1999), http://books.google.co.nz, accessed 9 October 2012, p. 34.
3 Ibid., p. 35.
4 Jean-Paul Lacroix, *Bois de Tonnellerie: De la forêt à la vigne et au vin* (2006), http://books.google.co.nz, accessed 15 September 2012, p. 57.
5 Jancis Robinson, ed., *The Oxford Companion to Wine* (Oxford, 1997), p. 276.
6 Kilby, *The Cooper and His Trade*, p. 131.
7 Ibid., p. 106.
8 Ibid., p. 112.
9 Ibid.
10 Ibid., p. 115.
11 Ibid.
12 Ibid., p. 116.
13 Ibid., p. 115.
14 Ibid., p. 116.
15 Ibid.
16 Ibid.
17 Ibid., p. 112.
18 Diana Twede, 'The Cask Age: The Technology and History of Wooden Barrels', *Packaging Technology and Science*, 18 (2005); and Kilby, *The Cooper and His Trade*.
19 Hunt and Murray, *A History of Business in Medieval Europe*, p. 35.
20 Garrett Oliver, *The Oxford Companion to Beer* (Oxford, 2012), p. 29.
21 Kilby, *The Cooper and His Trade*.
22 Jean Taransaud, *Le livre de LA TONNELLERIE* (Paris, 1976), p. 51; and Kilby, *The Cooper and His Trade*, p. 106.
23 George Elkington, *The Coopers: Company and Craft* (London, 1933), p. 63.
24 Ibid., p. 75.
25 Taransaud, *Le livre de LA TONNELLERIE*, p. 51; and Twede, 'The Cask Age', p. 257.

26  Kilby, *The Cooper and His Trade*, p. 106; and Taransaud, *Le livre de* LA TONNELLERIE, p. 51.

27  Taransaud, *Le livre de* LA TONNELLERIE, p. 207.

28  Kilby, *The Cooper and His Trade*, p. 122.

29  Elkington, *The Coopers*, p. 68.

30  The Worshipful Company of Coopers, www.coopers-hall.co.uk, accessed 10 January 2012.

31  Taransaud, *Le livre de* LA TONNELLERIE, p. 51.

32  Ibid.

33  Twede, 'The Cask Age', p. 258; and Kilby, *The Cooper and His Trade*, p. 120.

34  Kilby, *The Cooper and His Trade*, p. 120.

35  Ibid., p. 173.

36  Ibid.

37  Franklin E. Coyne, *The Development of the Cooperage Industry in the United States, 1620–1940* (Chicago, 1940), p. 17.

38  Ibid., p. 22.

39  Ibid., p. 19.

40  Kilby, *The Cooper and His Trade*, p. 163.

41  Coyne, *The Development of the Cooperage Industry*, p. 46.

42  Ibid.

43  Ibid., p. 69.

44  Ibid.

45  See archiver.rootsweb.ancestry.com, accessed 6 October 2012.

46  Coyne, *The Development of the Cooperage Industry*, p. 36.

47  The Associated Cooperage Industries of America, www.acia.net, accessed 7 October 2012.

48  Coyne, *The Development of the Cooperage Industry*, p. 34.

49  Lionel Mouraux and Jacques Champeix, *Bercy* (Paris, 1983), pp. 121–8.

50  Kilby, *The Cooper and His Trade*, p. 102.

51  Brian J. Cudahy, *Box Boats: How Container Ships Changed the World* (New York, 2006), p. xi.

### EIGHT
## Oak: Wood for Barrels

1 Jean Taransaud, *Le livre de* LA TONNELLERIE (Paris, 1976), p. 45.
2 Franklin E. Coyne, *The Development of the Cooperage Industry in the United States, 1620–1940* (Chicago, 1940), p. 36.
3 Howard A. Miller and Samuel H. Lamb, *Oaks of North America* (Happy Camp, CA, 1985), pp. 268–81.
4 See http://nrs.fs.fed.us, accessed 24 August 2010.
5 Miller and Lamb, *Oaks of North America*, p. 171.
6 Geoffrey Schahinger and Bryce Rankine, *Cooperage for Winemakers: A Manual on the Construction, Maintenance and Use of Oak Barrels* (Adelaide, 1999), pp. 26–8.
7 Coyne, *The Development of the Cooperage Industry*, p. 75.

### TEN
## Wine: Barrels and Oak Ageing

1 Kenneth Kilby, *The Cooper and His Trade* (Fresno, CA, 1977), p. 168.
2 Franklin E. Coyne, *The Development of the Cooperage Industry in the United States, 1620–1940* (Chicago, 1940), p. 27.
3 Howard A. Miller and Samuel H. Lamb, *Oaks of North America* (Happy Camp, CA, 1985), pp. 268–81.
4 Henry H. Work, 'Looking in All the Right Places: Eliminating TCA from Sources Other than Corks', *Practical Winery & Vineyard* (July/August 2004).
5 James Laube, 'Beaulieu Vineyard's Red Wine Woes', *The Wine Spectator*, XXVII/12 (2002), pp. 17–18.
6 Hugh Johnson, *The World Atlas of Wine* (New York, 1985), p. 134.
7 Henry H. Work, 'Innovation for the Winemaking Toolbox', *Practical Winery & Vineyard* (November/December 2007), p. 41.
8 Fiona Beckett, *Wine Uncorked* (London, 1999), p. 82.
9 Jancis Robinson, ed., *The Oxford Companion to Wine* (Oxford, 1997), pp. 99–100.
10 Vernon L. Singleton, Anthoula Randopoulo Sullivan and Cynthia Kramer, 'An Analysis of Wine to Indicate Aging in Wood or

Treatment with Wood Chips or Tannic Acid', *American Journal of Enology and Viticulture*, 22 (1971), pp. 161–6.

11 Robinson, *The Oxford Companion to Wine*, p. 100.

12 Ibid.

13 Allen Young, *Chardonnay, the World's Most Popular Grape: The Definitive Guide* (London, 1988), pp. 159–60.

14 Geoffrey Schahinger and Bryce Rankine, *Cooperage for Winemakers: A Manual on the Construction, Maintenance and Use of Oak Barrels* (Adelaide, 1999), p. 14.

## TWELVE
## Other Barrels: Spirits, Fortified Wines and Beer

1 Franklin E. Coyne, *The Development of the Cooperage Industry in the United States, 1620–1940* (Chicago, 1940), p. 11.

2 Ibid., p. 39.

3 Kenneth Kilby, *The Cooper and His Trade* (Fresno, CA, 1977), p. 48.

4 Janet Patton, 'Bourbon Barrels by the Millions', *Lexington Herald; Leader, Business Monday* (25 January 1999), p. 10.

5 Hugh Johnson, *The World Atlas of Wine* (New York, 1985), p. 300.

6 Coyne, *The Development of the Cooperage Industry*, p. 47.

7 Ibid.

8 Ibid., p. 98; and F. P. Hankerson, *The Wooden Barrel Manual of the Associated Cooperages of America* (St Louis, MO, 1951), pp. 38–43.

9 Johnson, *The World Atlas of Wine*, pp. 200–203.

10 Ibid., p. 291; and www.charlesnealselections.com, accessed 12 September 2012.

11 Hugh Johnson and Jancis Robinson, *The World Atlas of Wine*, 6th edn (London, 2007), p. 206.

12 Johnson, *The World Atlas of Wine*, p. 202.

13 Andrew Elder, *Whisky Map of Scotland* (Edinburgh, 1991).

14 Johnson, *The World Atlas of Wine*, p. 291.

15 Garrett Oliver, ed., *The Oxford Companion to Beer* (Oxford, 2012), p. 409.

16 Kenneth Kilby, *The Cooper and His Trade* (Fresno, CA, 1977), p. 91.

17 Ibid., pp. 160–61.

18 Oliver, *The Oxford Companion to Beer*, p. 448.

19 Ibid., pp. 96–8; and www.spbw.com, accessed 15 November 2012.
20 Oliver, *The Oxford Companion to Beer*, p. 98.
21 Ibid.
22 See www.budweiser.com, accessed 18 September 2012.
23 J. Gnagy, 'Beer Marketing Terms and What They Mean, Part Two: Beechwood Aging', http://thesoberbrewer.blogspot.co.nz, accessed 25 September 2012.
24 See www.nickelinstitute.org, accessed 19 January 2011.
25 See www.phrases.org.uk, accessed 18 September 2012.

## THIRTEEN
## Oak Flavouring: Oak Alternatives and Barrel Shaving

1 Diane O'Brien, 'Charles Shaw: Cheap Swills', www.brandchannel.com, 23 June 2003, accessed 15 November 2012.
2 Jancis Robinson, *The Oxford Companion to Wine*, 3rd edn (London, 2006), p. 491.
3 Henry H. Work, 'Oak Alternatives: The Next Generation', *Wine Technology* (October/November 2008).
4 Panos Kakaviatos, 'Oak Chips to Be Allowed in French Wine', www.decanter.com, 30 March 2006, accessed 20 November 2012.
5 Vaibhav, 'World of Wines and Wines of the World: Global Wine Consumption Trends', http://onvab.com/blog, accessed 18 November 2013.
6 '2012 Wine Sales in U.S. Reach New Record: Record California Winegrape Crop to Meet Surging Demand', www.wineinstitute.org, accessed 18 November 2013.

## FOURTEEN
## Cooperage: The Larger Picture

1 Joan Druett, *The Watery Grave* (New York, 2004), pp. 116–17.
2 See http://walmartstores.com, accessed 10 January 2012.
3 Ibid.

# Bibliography

Adkins, Leslie, and Roy Adkins, *Introduction to the Romans*
    (London, 1991)
Aubin, Gérard, Sandrine Lavaud and Philippe Roudié, *Bordeaux: Vignoble
    millénaire* (Bordeaux, 1996)
Beckett, Fiona, *Wine Uncorked* (London, 1999)
Blair, Peter H., *An Introduction to Anglo-Saxon England*, 3rd edn
    (Cambridge, 2003)
Bostock, John, and H. T. Riley, trans., *The Natural History of Pliny*, vol. III
    (1855), www.books.google.co.nz, accessed 1 September 2012
Canduci, Alexander, *Triumph and Tragedy: The Rise and Fall of Rome's
    Immortal Emperors* (Millers Point, NSW, 2010)
Chappell, William H. (c. 1855), personal diary while cooper on board the
    whaling ship *Saratoga*, retrieved via microfilm from the Alexander
    Turnbull Library, Wellington
Cook, James, *A Voyage towards the South Pole and round the World* (1777),
    http://books.google.co.nz, accessed 28 April 2011
Coyne, Franklin E., *The Development of the Cooperage Industry in the United
    States, 1620–1940* (Chicago, 1940)
Crawford, Anne, *A History of the Vintners' Company* (London, 1977)
Cudahy, Brian J., *Box Boats: How Container Ships Changed the World* (New
    York, 2006)
Cunliffe, Barry, ed., *Prehistoric Europe: An Illustrated History* (Oxford, 1996)
——, *The Ancient Celts* (Oxford, 1997)
Dando-Collins, Stephan, *Caesar's Legion: The Epic Saga of Julius Caesar's Elite
    Tenth Legion and the Armies of Rome* (New York, 2002)

Davies, Norman, *The Isles* (London, 1999)

Davis, John, *The Celts* (London, 2000)

Dean, Bob, 'A 3rd Century AD Gallo-Roman Trading Vessel from Guernsey aka The St. Peter Port Wreck or "The Asterix Ship"', www.swan.ac.uk, accessed 26 September 2012

Druett, Joan, *A Watery Grave* (New York, 2004)

Eco, Umberto, *Serendipities: Language and Lunacy*, trans. William Weaver (London, 1998)

Elder, Andrew, *Whisky Map of Scotland* (Edinburgh, 1991)

Elkington, George, *The Coopers: Company and Craft* (London, 1933)

Ellis, Peter Berresford, *The Ancient World of the Celts* (London, 1998)

Fagan, Brian, *The Long Summer: How Climate Changed Civilization* (New York, 2004)

Fenton, James, ed., *Samuel Taylor Coleridge: Poems* (London, 2011)

Gascoigne, John, *Captain Cook: Voyager between Worlds* (London, 2007)

Gnagy, J., 'Beer Marketing Terms and What They Mean, Part Two: Beechwood Aging', http://thesoberbrewer.blogspot.co.nz, accessed 25 September 2007

Green, Jen, Fiona MacDonald, Philip Steele and Michael Stotter, *The Encyclopedia of the Ancient Americas* (London, 2000)

Guzzwell, John, *Modern Wooden Yacht Construction: Cold-molding, Joinery, Fitting Out* (London, 1979)

Hankerson, F. P., *The Wooden Barrel Manual of the Associated Cooperages of America* (St Louis, MO, 1951)

Hunt, Edwin S., and James M. Murray, *A History of Business in Medieval Europe, 1200–1550* (1999), http://books.google.co.nz, accessed 9 October 2012

James, Margery K., 'The Medieval Wine Dealer', *Explorations in Entrepreneurial History*, X/2 (1957), pp. 45–53

James, Simon, *Exploring the World of the Celts* (London, 1993)

Johnson, Hugh, *The World Atlas of Wine* (New York, 1985)

——, and Jancis Robinson, *The World Atlas of Wine*, 6th edn (London, 2007)

Kakaviatos, Panos, 'Oak Chips to Be Allowed in French Wine', www.decanter.com, 30 March 2006

Kennedy, Jennifer, 'Brief History of Cod Fishing', http://marinelife.about.com, 15 May 2011

Kilby, Kenneth, *The Cooper and His Trade* (Fresno, CA, 1977)

Lacroix, Jean-Paul, Bois de Tonnellerie: De la forêt à la vigne et au vin (2006),
    http://books.google.co.nz, accessed 15 September 2012

Lambert, Ron, and Gail Henry, Taranaki: An Illustrated History
    (Auckland, 2000)

Laube, James, 'Beaulieu Vineyard's Red Wine Woes', The Wine Spectator,
    XXVII/12 (2002), pp. 17–18

Liversidge, Joan, Britain in the Roman Empire (London, 1969)

Logan, William Bryant, Oak: The Frame of Civilization (New York, 2005)

MacKenzie, Jr, Clyde L., 'History of Oystering in the United States and
    Canada, Featuring the Eight Greatest Oysters', Marine Fisheries Review,
    LVIII/4 (1996), pp. 1–78

Marrow, Josh, '2003 San Simeon Earthquake, Lessons Learned on the
    Fault Line', Practical Winery and Vineyard (2004, May–June)

Mawer, Granville Allen, Ahab's Trade: The Saga of South Seas Whaling
    (Leonards, NSW, 1999)

McDevitte, W. A., and W. S. Bohn, trans., The Works of Julius Caesar,
    www.sacred-texts.com, accessed 3 October 2012

Menzel, Donald H., A Field Guide to the Stars and Planets (Boston, MA, 1964)

Miller, Howard A., and Samuel H. Lamb, Oaks of North America (Happy
    Camp, CA, 1985)

Morris, Ian, Why the West Rules – For Now (London, 2010)

Morton, Harry, The Wind Commands: Sailors and Sailing Ships in the Pacific
    (Dunedin, New Zealand, 1975)

Mouraux, Lionel, and Jacques Champeix, Bercy (Paris, 1983)

Nares, George S., Seamanship, 2nd edn (Facsimile edn, Woking, 1979
    [1862])

O'Brien, Diane, 'Charles Shaw: Cheap Swills', www.brandchannel.com,
    23 June 2003

Oliver, Garrett, ed., The Oxford Companion to Beer (Oxford, 2012)

Patton, Janet, 'Bourbon Barrels by the Million', Lexington Herald; Leader,
    Business Monday (25 January 1999), pp. 10–11

Phillips, Charles, and D. M. Jones, The Everyday Life of the Aztec and Maya
    (London, 2007)

Pitt, Catherine Rachel, 'The Wine Trade in Bristol in the Fifteenth and
    Sixteenth Centuries', www.bris.ac.uk, accessed 20 October 2012

Robinson, Jancis, ed., The Oxford Companion to Wine (Oxford, 1997)

——, ed. The Oxford Companion to Wine, 3rd edn (Oxford, 2006)

Schahinger, Geoffrey, and Bryce Rankine, *Cooperage for Winemakers: A Manual on the Construction, Maintenance and Use of Oak Barrels* (Adelaide, 1999)

Singleton, Vernon L., Anthoula Randopoulo Sullivan and Cynthia Kramer, 'An Analysis of Wine to Indicate Aging in Wood or Treatment with Wood Chips or Tannic Acid', *American Journal of Enology and Viticulture*, 22 (1971), pp. 161–6

Soyez, Jean-Marc, *Les Ebenistes du Vin*, trans. Michael Mills (Lormont, France, 1991)

Spectre, Peter H., and Maynard Bray, eds, *The Wooden Boat Series: Planking and Fastening* (Brooklin, MA, 1996)

St J. de Crèvecoeur, John H., *Letters from an American Farmer* (1782), http://xroad.virginia.edu, accessed 9 May 2011

Taransaud, Jean, *Le livre de LA TONNELLERIE* (Paris, 1976)

Thomas, Nicholas, *Cook: The Extraordinary Voyages of Captain James Cook* (New York, 2003)

Thrupp, Sylvia L., *The Merchant Class of Medieval London* (1988 [1948]), http://books.google.co.nz, accessed 9 October 2012

Twede, Diana, 'Commercial Amphoras: The Earliest Consumer Packages?', *Journal of Macromarketing*, 22 (2002), pp. 98–108

——, 'The Cask Age: The Technology and History of Wooden Barrels', *Packaging Technology and Science*, 18 (2005), pp. 253–65

Tylecote, R. F., *The History of Metallurgy* (London, 1976)

Work, Henry H., 'Looking in All the Right Places: Eliminating TCA from Sources Other than Corks', *Practical Winery & Vineyard* (July/August 2004)

——, 'Innovation for the Winemaking Toolbox', *Practical Winery & Vineyard* (November/December 2007), pp. 34, 36–8, 40–42, 44–6

——, 'Oak Alternatives: The Next Generation', *Wine Technology* (October/November 2008)

Young, Allen, *Chardonnay, The World's Most Popular Grape: The Definitive Guide* (London, 1988)

# Acknowledgements

Many thanks to my wife, Karen, who has supported and tolerated my quest for cooperage knowledge for countless years; to Jack Kearns, a long-time friend, for his diligent editing and suggestions; to Corey Mindlin for her intelligent thoughts and ideas concerning the story line and, early on in the process, to Stuart Work for his suggestions. A special thank you to my close cooperage associates: Phil Burton, Ken Seymour and those associated with Canton Cooperage, especially Bill Weil, Bill Newton, Fay Beech and J. D. Ray; to the good people at Taransaud, Nadalie, Tonnellerie de Bourgogne, A&K and all the other cooperages and coopers I have had the pleasure of knowing and working with. Thanks to Don Neel, editor at *Practical Winery*, for publishing a number of my articles – from which I have drawn some of the information in the book. Thanks to all those who have been patient with my continuous ramblings about barrels. And, finally, thanks to my Reaktion editors, Ben Hayes and Martha Jay, who skilfully and patiently guided me through the publishing process.

# Acknowledgements

# Photo Acknowledgements

Courtesy Gilles Arroyo: p. 21; courtesy of Barrel Builders, St Helena, California, and *Practical Winery & Vineyard*, San Rafael, California, USA: p. 201; © Bibliothèque nationale de France, Paris: pp. 26, 40, 78, 79, 89, 120, 143; © Trustees of the British Museum, London: pp. 6, 13, 52, 65, 76, 107, 151, 160, 182; courtesy Conrad Cichorius: p. 55; courtesy of the Drake Well Museum, Pennsylvania Historical Museum Commission, Titusville, Pennsylvania, USA: p. 116; Dreamstime LLC: pp. 30 (Steve Estvanik), 175 (Chris Hellyar), 185 (Olikli), 190 (Volgariver), 196 (Toniflap), 208 (Mgs99); Jamie Goode: p. 48; Robert Farrow, courtesy of the Guernsey Museums and Galleries (States of Guernsey), St Peter Port, Guernsey: p. 62; Hunterian Gallery of Art, University of Glasgow, Scotland: p. 109; Library of Congress, Prints and Photographs Division, FSA/OWI Collection: p. 206; courtesy of Hans Muthmann: pp. 18, 19, 24, 49, 92, 127, 129, 130, 135, 136, 169; courtesy of the New Bedford Whale Museum, New Bedford, Massachusetts, USA: pp. 70–71, 102; courtesy of Oak Solutions Group of Aftek Filters, Napa, California, and *Practical Winery & Vineyard*, San Rafael, California: p. 200; courtesy Mark Peters: p. 14; S. Victor White, © Reading Museum (Reading Borough Council), all rights reserved. Reading, UK: p. 37; Thomas Zühmer, Rheinisches Landesmuseum, Trier, Germany: p. 56; courtesy of Seppeltsfield Winery, Barossa Valley, South Australia: p. 178; courtesy of Tonnellerie Nadalié, Bordeaux, France: pp. 144 top, 171; courtesy of Twomey Cellars and Silver Oak Winery, Oakville, California, USA: p. 157; Henry H. Work: p. 139; Karen Work: p. 158; copyright © by kind permission of The Worshipful Company of Coopers, London: p. 113; courtesy of www.sherry.org, Jerez de la Frontera (Cádiz), Spain: pp. 144 bottom, 193.

# Index